P9-DNW-109

CQR Pocket Guide
Anatomy and Physiology

The Anatomy of a Cell

The cell is the basic functional unit of all living things. The plasma membrane of the cell has the following functions:

- Bounds the cell and encloses the nucleus and cytoplasm.
- Separates internal metabolic events from the external environment.
- Controls the movement of materials into and out of the cell.

Organelles are bodies within the cytoplasm that serve to physically separate the various metabolic reactions that occur within cells. They include the following:

- Nucleus
- Endoplasmic reticulum (ER)
- Golgi apparatus
- Lysosomes
- Mitochondria
- Microtubules, intermediate filaments, and microfilaments
- Flagella and cillia
- Centrioles and basal bodies
- Peroxisomes

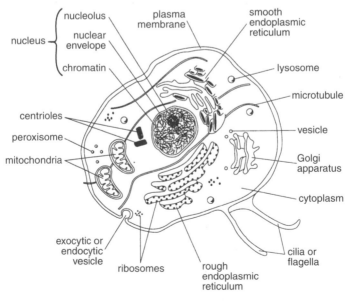

CQR Pocket Guide
Anatomy and Physiology

The Skeletal System

The human body contains 206 bones. The bones of the body are categorized into two groups:

- **The axial skeleton:** The bones of the axial skeleton revolve around the vertical axis of the skeleton, basically your spine and all the bones of your skull.

- **The appendicular skeleton:** The bones of the appendicular skeleton make up the limbs that have been appended to the axial skeleton, basically your arms, legs, hands, and feet.

When references are made to parts of the body, they refer to the body in the anatomical position—standing erect, facing forward, arms down at the side, with the palms turned forward.

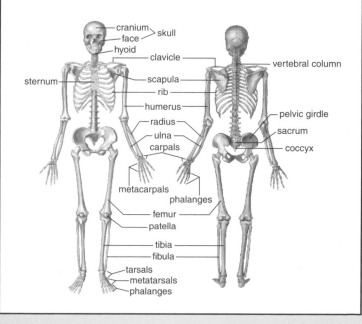

Copyright © 2001 Wiley Publishing, Inc
All rights reserved.
For more information about Wiley Publishing,
call 1-800-762-2974.

Wiley, the Wiley Publishing logo, For Dummies, the Dummies Man logo, the For Dummies Bestselling Book Series logo and all related trade dress are trademarks or registered trademarks of Wiley Publishing, Inc. All other trademarks are property of their respective owners.

CliffsQuickReview™
Anatomy and
Physiology

By Phillip E. Pack, Ph.D.

Wiley Publishing, Inc.

About the Author

Phillip E. Pack has taught AP courses and gifted programs for 11 years. He is currently an assistant professor of math and science at Woodbury University in Burbank, California.

Publisher's Acknowledgments

Editorial

Project Editor: Kelly Ewing

Acquisitions Editor: Sherry Gomoll

Editorial Coordinator: Michelle Hacker

General Reviewer: Mark F. Seifert, PhD

Special Help: Donna Frederick

Composition

Indexer: TECHBOOKS Production Services

Proofreader: TECHBOOKS Production Services

Wiley Publishing Indianapolis Composition Services

CliffsQuickReview™ Anatomy & Physiology

Published by

Wiley Publishing, Inc.

111 River Street

Hoboken, NJ 07030-5774

www.wiley.com

Note: If you purchased this book without a cover you should be aware that this book is stolen property. It was reported as "unsold and destroyed" to the publisher, and neither the author nor the publisher has received any payment for this "stripped book."

Copyright © 2001 Wiley Publishing, Inc. New York, New York

Library of Congress Control Number: 2001024133

ISBN: 978-0-7645-6373-7

Printed in the United States of America

15 14 13

1O/TR/QT/QY/IN

No part of this publication may be reproduced, stored in a retrieval system, or transmitted in any form or by any means, electronic, mechanical, photocopying, recording, scanning, or otherwise, except as permitted under Sections 107 or 108 of the 1976 United States Copyright Act, without either the prior written permission of the Publisher, or authorization through payment of the appropriate per-copy fee to the Copyright Clearance Center, 222 Rosewood Drive, Danvers, MA 01923, 978-750-8400, fax 978-750-4744. Requests to the Publisher for permission should be addressed to the Legal Department, Wiley Publishing, Inc., 10475 Crosspoint Blvd., Indianapolis, IN 46256, 317-572-3447, fax 317-572-4447, or e-mail permcoordinator@wiley.com

LIMIT OF LIABILITY/DISCLAIMER OF WARRANTY: THE PUBLISHER AND AUTHOR HAVE USED THEIR BEST EFFORTS IN PREPARING THIS BOOK. THE PUBLISHER AND AUTHOR MAKE NO REPRESENTATIONS OR WARRANTIES WITH RESPECT TO THE ACCURACY OR COMPLETENESS OF THE CONTENTS OF THIS BOOK AND SPECIFICALLY DISCLAIM ANY IMPLIED WARRANTIES OF MERCHANTABILITY OR FITNESS FOR A PARTICULAR PURPOSE. THERE ARE NO WARRANTIES WHICH EXTEND BEYOND THE DESCRIPTIONS CONTAINED IN THIS PARAGRAPH. NO WARRANTY MAY BE CREATED OR EXTENDED BY SALES REPRESENTATIVES OR WRITTEN SALES MATERIALS. THE ACCURACY AND COMPLETENESS OF THE INFORMATION PROVIDED HEREIN AND THE OPINIONS STATED HEREIN ARE NOT GUARANTEED OR WARRANTED TO PRODUCE ANY PARTICULAR RESULTS, AND THE ADVICE AND STRATEGIES CONTAINED HEREIN MAY NOT BE SUITABLE FOR EVERY INDIVIDUAL. NEITHER THE PUBLISHER NOR AUTHOR SHALL BE LIABLE FOR ANY LOSS OF PROFIT OR ANY OTHER COMMERCIAL DAMAGES, INCLUDING BUT NOT LIMITED TO SPECIAL, INCIDENTAL, CONSEQUENTIAL, OR OTHER DAMAGES. FULFILLMENT OF EACH COUPON OFFER IS THE RESPONSIBILITY OF THE OFFEROR.

Trademarks: Wiley, the Wiley Publishing logo, Cliffs, CliffsNotes, the CliffsNotes logo, CliffsAP, CliffsComplete, CliffsTestPrep, CliffsQuickReview, CliffsNote-a-Day and all related logos and trade dress are registered trademarks or trademarks of Wiley Publishing, Inc., in the United States and other countries. All other trademarks are property of their respective owners. Wiley Publishing, Inc., is not associated with any product or vendor mentioned in this book.

For general information on our other products and services or to obtain technical support, please contact our Customer Care Department within the U.S. at 800-762-2974, outside the U.S. at 317-572-3993, or fax 317-572-4002.

Wiley also publishes its books in a variety of electronic formats. Some content that appears in print may not be available in electronic books.

Table of Contents

INTRODUCTION

Everyone from high school students to medical students needs to have a basic understanding of human anatomy and physiology. If you understand how your body is built and the different functions it performs, you may appreciate it more than you probably do.

The human body is complex and houses many systems. A general understanding of biology is helpful in understanding anatomy, but not necessary. A general knowledge of chemistry is beneficial in understanding physiology. Don't worry if you don't have that knowledge; this book gives you the basics so you can understand the rest.

Why You Need This Book

Can you answer yes to any of these questions?

- Do you need to review the fundamentals of anatomy and physiology fast?

- Do you need a course supplement to human anatomy and physiology?

- Do you need a concise, comprehensive reference for anatomy and physiology?

If so, then CliffsQuickReview Anatomy & Physiology is for you!

How to Use This Book

You can use this book in any way that fits your personal style for study and review—you decide what works best with your needs. You can either read the book from cover to cover or just look for the information you want and put it back on the shelf for later. Here are just a few ways you can search for topics:

- Use the Pocket Guide to find essential information, such as the anatomy of a cell and the major bones of the human skeletal system.

- Look for areas of interest in the book's Table of Contents, or use the index to find specific topics.

■ Flip through the book looking for subject areas at the top of each page.

■ Get a glimpse of what you'll gain from a chapter by reading through the "Chapter Check-In" at the beginning of each chapter.

■ Use the Chapter Checkout at the end of each chapter to gauge your grasp of the important information you need to know.

■ Test your knowledge more completely in the CQR Review and look for additional sources of information in the CQR Resource Center.

■ Use the glossary to find key terms fast. This book defines new terms and concepts where they first appear in the chapter. If a word is bold-faced, you can find a more complete definition in the book's glossary.

■ Or flip through the book until you find what you're looking for—we organized this book to gradually build on key concepts.

Visit Our Web Site

A great resource, www.cliffsnotes.com, features review materials, valuable Internet links, quizzes, and more to enhance your learning. The site also features timely articles and tips, plus downloadable versions of many CliffsNotes books.

When you stop by our site, don't hesitate to share your thoughts about this book or any Wiley product. Just click the Talk to Us button. We welcome your feedback!

Chapter 1

ANATOMY AND CHEMISTRY BASICS

Chapter Check-In

❑ Understanding the basics of anatomy

❑ Noting the basic chemical constituents that help form matter

❑ Listing the types of bonds that form between two atoms

❑ Understanding the difference between inorganic and organic compounds

❑ Describing the four classes of organic molecules

❑ Finding out how a chemical reaction occurs in a biological system

After you know the basic terms of anatomy but before studying the structure and function of the body, you need to have a basic knowledge of chemistry that will be pertinent to your studies. Some of the chemistry presented in this chapter may not be new to you. In fact, the organic molecules of carbohydrates, lipids (fats), and proteins are the staple of a healthy diet and lifestyle. Learning these basic chemical components is essential for future studies in physiology, nutrition, and many other fields of scientific interest.

So What Is Anatomy?

Anatomy is the study of the structure and relationship between body parts. **Physiology** is the study of the function of body parts and the body as a whole. Some specializations within each of these sciences follow.

■ **Gross (macroscopic) anatomy** is the study of body parts visible to the naked eye, such as the heart or bones.

- Histology is the study of tissues at the microscopic level.

- **Cytology** is the study of cells at the microscopic level.

- Neurophysiology is the study of how the nervous system functions.

Organizations of living systems

Living systems can be defined from various perspectives, from the very broad (looking at the entire earth) to the very minute (individual atoms). Each perspective provides information about how or why a living system functions.

- At the chemical level, atoms, molecules (combinations of atoms), and the chemical bonds between atoms provide the framework upon which all living activity is based.

- The cell is the smallest unit of life. Organelles within the cell are specialized bodies performing specific cellular functions. Cells themselves may be specialized. Thus, there are nerve cells, bone cells, and muscle cells.

- A tissue is a group of similar cells performing a common function. Muscle tissue, for example, consists of muscle cells.

- An **organ** is a group of different kinds of tissues working together to perform a particular activity. The heart is an organ composed of muscle, nerve, connective, and epithelial tissues.

- An organ system is two or more organs working together to accomplish a particular task. The digestive system, for example, involves the coordinated activities of many organs, including the mouth, stomach, small and large intestines, pancreas, and liver.

- An **organism** is a system possessing the characteristics of living things—the ability to obtain and process energy, the ability to respond to environmental changes, and the ability to reproduce.

Homeostasis

A characteristic of all living systems is **homeostasis,** or the maintenance of stable, internal conditions within specific limits. In many cases, stable conditions are maintained by negative feedback.

In negative feedback, a sensing mechanism (a receptor) detects a change in conditions beyond specific limits. A control center, or **integrator** (often the brain), evaluates the change and activates a second mechanism (an effector) to correct the condition. Conditions are constantly monitored by receptors and evaluated by the control center. When the control center determines that conditions have returned to normal, corrective action is discontinued. Thus, in negative feedback, the variant condition is canceled, or negated, so that conditions are returned to normal. Compare this with positive feedback, in which an action intensifies a condition so that it is driven farther beyond normal limits. Such positive feedback is uncommon but does occur during childbirth (labor contractions), lactation (where milk production increases in response to an increase in nursing), and sexual orgasm.

The regulation of glucose concentration in the blood illustrates how homeostasis is maintained by negative feedback. After a meal, the absorption of glucose (a sugar) from the digestive tract increases the amount of glucose in the blood. In response, specialized cells in the pancreas secrete the hormone insulin, which circulates through the blood and stimulates liver and muscle cells to absorb the glucose. Once blood glucose levels return to normal, insulin secretion stops. Later, perhaps after heavy exercise, blood glucose levels may drop because muscle cells absorb glucose from the blood and use it as a source of energy for muscle contraction. In response to falling blood glucose levels, another group of specialized pancreatic cells secretes a second hormone, glucagon. Glucagon stimulates the liver to release its stored glucose into the blood. When blood glucose levels return to normal, glucagon secretion stops.

Anatomic terminology

In order to identify areas of the body accurately, clearly defined anatomical terms are used. These terms refer to the body in the anatomical position—standing erect, facing forward, arms down at the side, with the palms turned forward. In this position, the following terms apply.

■ Directional terms are used to describe the relative position of one body part to another. These terms are listed in Table 1-1.

■ Body planes and sections are used to describe how the body or an organ is divided into two parts.

> **Sagittal planes** divide a body or organ vertically into a right and left part. If the right and left parts are equal, the plane is the midsagittal plane; if they're unequal, the plane is a parasagittal plane.

> A **frontal (coronal) plane** divides the body or organ vertically into a front and rear part.

A **horizontal (transverse) plane** divides the body or organ horizontally into a top and bottom part.

■ Body cavities are enclosed areas that house organs. These cavities are organized into two groups:

The **dorsal** (back) body cavity includes the cranial cavity (which contains the brain) and the vertebral cavity (which contains the spinal cord).

The ventral (front) body cavity includes the thoracic cavity (which contains the lungs, each in its own pleural cavity, and the heart, in the pericardial cavity) and the abdominopelvic cavity (which contains the digestive organs in the abdominal cavity and the bladder and reproductive organs in the pelvic cavity).

■ Regional terms identify specific areas of the body. In some cases, a descriptive word is used to identify the location. For example, the axial region refers to the main axis of the body—the head, neck, and trunk. The appendicular region refers to the appendages—the arms and legs. Other regional terms use a body part to identify a particular region of the body. For example, the nasal region refers to the nose.

Table 1-1   Basic Anatomy Terms

Term	Definition	Example
Superior	Above another structure.	The heart is superior to the stomach.
Inferior	Below another structure.	The stomach is inferior to the heart.
Anterior (or ventral)	Toward the front of the body.	The navel is anterior to the spine.
Posterior (or dorsal)	Toward the back of the body.	The spine is posterior to the navel.
Medial	Toward the midline of the body. (The midline divides the body into equal right and left sides.)	The nose is medial to the eyes.
Lateral	Away from the midline of the body (or toward the side of the body).	The ears are lateral to the nose.
Ipsilateral	On the same side of the body.	The spleen and descending colon are ipsilateral.
Contralateral	On opposite sides of the body.	The ascending and descending colons are contralateral.

Term	Definition	Example
Intermediate	Between two structures.	The knee is intermediate between the thigh and leg.
Proximal	Closer to the point of attachment of a limb.	The elbow is proximal to the wrist.
Distal	Father from the point of attachment of a limb.	The foot is distal to the knee.
Superficial	Toward the surface of the body.	The skin is superficial to the muscle.
Deep	Away from the surface of the body.	The skeleton is deep to the skin.

Atoms, Molecules, Ions, and Bonds

Matter is anything that takes up space and has mass. Matter consists of **elements** that possess unique physical and chemical properties. Elements are represented by chemical symbols of one or two letters, such as C (carbon), Ca (calcium), H (hydrogen), O (oxygen), N (nitrogen), and P (phosphorus). The smallest quantity of an element that still possesses the characteristics of that element is an **atom.** Atoms chemically bond together to form **molecules,** and the composition of a molecule is given by its chemical formula (O_2, H_2O, $C_6H_{12}O_6$). When the atoms in a molecule are different, the molecule is a **compound** (H_2O and $C_6H_{12}O_6$, but not O_2).

The atoms of every element consist of a nucleus of positively charged **protons** and neutrally charged **neutrons.** Negatively charged electrons are arranged outside the nucleus. The atoms of each element differ by their number of protons, neutrons, and **electrons.** For example, hydrogen has one proton, one electron, and no neutrons, while carbon has six protons, six neutrons, and six electrons. The number and arrangement of electrons of an atom determine the kinds of chemical bonds that it forms and how it reacts with other atoms to form molecules. There are three kinds of chemical bonds.

■ Ionic bonds form between two atoms when one or more electrons are completely transferred from one atom to the other. The atom that gains electrons has an overall negative charge, and the atom that donates electrons has an overall positive charge. Because of their positive or negative charges, these atoms are ions. The attraction of the positive ion to the negative ion constitutes the ionic bond. Sodium (Na) and chlorine (Cl) form ions (Na+ and Cl-), which attract one another to form the ionic bond in a sodium chloride (NaCl) molecule.

A plus or minus sign following a chemical symbol indicates an ion with a positive or negative charge that results from the loss or gain of one or more electrons, respectively. Numbers preceding the charges indicate ions whose charges are greater than one (Ca^{2+}, PO_{32-}).

■ **Covalent** bonds form when electrons are shared between atoms. That is, neither atom completely retains possession of the electrons (as happens with atoms that form ionic bonds). A single covalent bond occurs when two electrons are shared (one from each atom). A double or triple covalent bond is formed when four or six electrons are shared, respectively. When the two atoms sharing electrons are exactly the same, as in a molecule of oxygen gas (two oxygen atoms to form O_2), the electrons are shared equally, and the bond is a nonpolar covalent bond. When the atoms are different, such as in a molecule of water (H_2O), the larger nucleus of the oxygen atom exerts a stronger pull on the shared electrons than does the single proton that makes up either hydrogen nucleus. In this case, a polar covalent bond is formed because the unequal distribution of the electrons creates areas within the molecule that have either a negative or positive charge (or pole), as shown in Figure 1-1.

■ **Hydrogen bonds** are weak bonds that form between the positively charged hydrogen atom in one covalently bonded molecule and the negatively charged area of another covalently bonded molecule. Water, for example, forms hydrogen bonds between water molecules. Since the atoms in water form a polar covalent bond, the positive area in H_2O around the hydrogen proton attracts the negative areas in an adjacent H_2O molecule. This attraction forms the hydrogen bond (Figure 1-1b).

Figure 1-1   Two examples of chemical bonds.

(a) A Water Molecule Showing Polarity Created by Covalent Bonds   **(b) Hydrogen Bonding Between Water Molecules**

Inorganic Compounds

Inorganic compounds are typically compounds without carbon atoms. Water, O_2, and NaCl are examples of inorganic compounds.

Water is the most abundant substance in the body. Its abundance is due partly to its unique chemical properties created by the influence of its hydrogen bonds. These properties include the following:

■ Water is an excellent solvent. Ionic substances are soluble in water (they dissolve) because the poles of the polar water molecules pull them apart, forming ions. Polar covalent substances are also water-soluble because they share the same hydrogen bonding as water shares with itself. For this reason, polar covalent substances are called hydrophilic (water loving). Because they lack charged poles, nonpolar covalent substances do not dissolve in water and are called hydrophobic (water fearing).

■ Because water molecules are held together by hydrogen bonds, water molecules have a high degree of cohesion, or the ability to stick together. As a result, water has strong surface tension. This tension, in turn, gives water strong capillary action, allowing water to creep up narrow tubing. These qualities contribute to the movement of water through animal capillaries.

■ The temperature of water is very stable. You must add a relatively large amount of energy to warm (and boil) it and remove a large amount of energy to cool (and freeze) it. So when sweat evaporates from your forehead, a large amount of heat is taken with it and you are cooled.

Organic Molecules

Organic compounds are those that have carbon atoms. In living systems, large organic molecules, called macromolecules, may consist of hundreds or thousands of atoms. Most macromolecules are polymers, molecules that consist of a single unit (monomer) repeated many times.

Four of carbon's six electrons are available to form bonds with other atoms. Thus, you will always see four lines connecting a carbon atom to other atoms, each line representing a pair of shared electrons (one electron from carbon and one from another atom). Complex molecules can be formed by stringing carbon atoms together in a straight line or by connecting carbons together to form rings. The presence of nitrogen, oxygen, and other atoms adds additional variety to these carbon molecules.

Four important classes of organic molecules—carbohydrates, lipids, proteins, and nucleic acids—are discussed in the following sections.

Carbohydrates

Carbohydrates are classified into three groups according to the number of sugar (or saccharide) molecules present.

■ A **monosaccharide** is the simplest kind of carbohydrate. It is a single sugar molecule, such as a fructose or glucose (Figure 1-2). Sugar molecules have the formula $(CH_2O)_n$, where n is any number from 3 to 8. For glucose, n is 6, and its formula is $C_6H_{12}O_6$. The formula for fructose is also $C_6H_{12}O_6$, but as you can see in Figure 1-2, the placement of the carbon atoms is different. Very small changes in the position of certain atoms, such as those that distinguish glucose and fructose, may dramatically change the chemistry of a molecule.

■ A **disaccharide** consists of two linked sugar molecules. Glucose and fructose, for example, link to form sucrose (see Figure 1-2).

■ A **polysaccharide** consists of a series of connected monosaccharides. Thus, a polysaccharide is a polymer because it consists of repeating units of monosaccharide. Starch is a polysaccharide made up of a thousand or more glucose molecules and is used in plants for energy storage. A similar polysaccharide, glycogen, is used in animals for the same purpose.

Lipids

Lipids are a class of substances that are insoluble in water (and other polar solvents), but are soluble in nonpolar substances (like ether or chloroform). There are three major groups of lipids.

■ Triglycerides include fats, oils, and waxes. They consist of three fatty acids attached to a glycerol molecule (Figure 1-3). Fatty acids are hydrocarbons (chains of covalently bonded carbons and hydrogens) with a carboxyl group (-COOH) at one end of the chain. A saturated fatty acid has a single covalent bond between each pair of carbon atoms, and each carbon has two hydrogens bonded to it. You can remember this fact by thinking that each carbon is "saturated" with hydrogen. An unsaturated fatty acid occurs when a double covalent bond replaces a single covalent bond and two hydrogen atoms (Figure 1-3). Polyunsaturated fatty acids have many of these double bonds.

Figure 1-2 Molecular structure of several carbohydrates.

glucose

fructose

sucrose

Figure 1-3 The molecular structure of a triglyceride.

glycerol + 3 fatty acids = triglyceride

saturated fatty acids

unsaturated fatty acids

A Triglyceride

■ Phospholipids look just like lipids except that one of the fatty acid chains is replaced by a phosphate ($-PO_{32-}$) group (Figure 1-4). Additional chemical groups (indicated by R in Figure 1-4) are usually attached to the phosphate group. Since the fatty acid "tails" of phospholipids are nonpolar and hydrophobic and the glycerol and phosphate "heads" are polar and hydrophilic, phospholipids are often found oriented in sandwich-like formations with the hydrophilic heads oriented toward the outside. Such formations of phospholipids provide the structural foundation of cell membranes.

Figure 1-4 The molecular structure of a phospholipid.

A Phospholipid

■ **Steroids** are characterized by a backbone of four linked carbon rings (Figure 1-5). Examples of steroids include cholesterol (a component of cell membranes) and certain hormones, including testosterone and estrogen.

Figure 1-5 Examples of steroids.

steroid backbone

testosterone

cholesterol

Proteins

Proteins represent a class of molecules that have varied functions. Eggs, muscles, antibodies, silk, fingernails, and many hormones are partially or entirely proteins. Although the functions of proteins are diverse, their structures are similar. All proteins are polymers of amino acids; that is, they consist of a chain of amino acids covalently bonded. The bonds between the amino acids are called peptide bonds, and the chain is a polypeptide, or peptide. One protein differs from another by the number of arrangement of the 20 different amino acids. Each amino acid consists of a central carbon bonded to an amino group ($-NH_2$), a carboxyl group ($-COOH$), and a hydrogen atom (Figure 1-6). The fourth bond of the central carbon is shown with the letter R, which indicates an atom or group of atoms that varies from one kind of amino acid to another. For the simplest amino acid, glycine, the R is a hydrogen atom. For serine, R is CH_2OH. For other amino acids, R may contain sulfur (as in cysteine) or a carbon ring (as in phenyalanine).

Figure 1-6 Examples of amino acids.

There are four levels that describe the structure of a protein:

■ The primary structure of a protein describes the order of amino acids. Using three letters to represent each amino acid, the primary structure for the protein antidiuretic hormone (ADH) can be written as cys-tyr-glu-asn-cys-pro-arg-gly.

- The secondary structure of a protein is a three-dimensional shape that results from hydrogen bonding between amino acids. The bonding produces a spiral (alpha helix) or a folded plane that looks much like the pleats on a skirt (beta pleated sheet).

- The tertiary structure of a protein includes additional three-dimensional shaping that results from interaction among R groups. For example, hydrophobic R groups tend to clump toward the inside of the protein, while hydrophilic R groups clump toward the outside of the protein. Additional three-dimensional shaping occurs when the amino acid cysteine bonds to another cysteine across a disulfide bond. This causes the protein to twist around the bond (Figure 1-7).

Figure 1-7 Disulfide bonds can dictate a protein's structure.

- The quaternary structure describes a protein that is assembled from two or more separate peptide chains. The protein hemoglobin, for example, consists of four peptide chains that are held together by hydrogen bonding, interactions among R groups, and disulfide bonds.

Nucleic acids

The genetic information of a cell is stored in molecules of **deoxyribonucleic acid (DNA).** The DNA, in turn, passes its genetic instructions to ribonucleic acid (RNA) for directing various metabolic activities of the cell.

DNA is a polymer of nucleotides (Figure 1-8). A DNA molecule consists of three parts—a nitrogen base, a five-carbon sugar called deoxyribose, and a phosphate group. There are four DNA nucleotides, each with one of the four nitrogen bases (adenine, thymine, cytosine, and guanine). The first letter of each of these four bases is often used to symbolize the respective nucleotide (A for adenine nucleotide, for example).

Figure 1-8 The molecular structure of nucleotides.

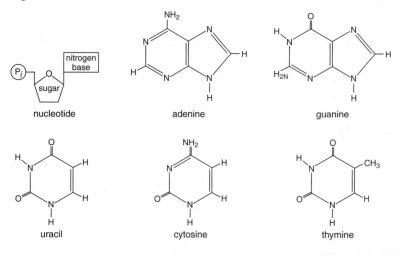

nucleotide

adenine

guanine

uracil

cytosine

thymine

Figure 1-9 shows how two strands of nucleotides, paired by weak hydrogen bonds between the bases, from a double-stranded DNA. When bonded in this way, DNA forms a two-stranded spiral, or double helix. Note that adenine always bonds with thymine and cytosine always bonds with guanine.

RNA differs from DNA in the following ways:

- The sugar in the nucleotides that make an RNA molecule is ribose, not deoxyribose as it is in DNA.

- The thymine nucleotide does not occur in RNA. It is replaced by uracil. When pairing of bases occurs in RNA, uracil (instead of thymine) pairs with adenine.

- RNA is usually single-stranded and does not form a double helix as does DNA.

Figure 1-9 Two-dimensional illustrations of the structure of DNA.

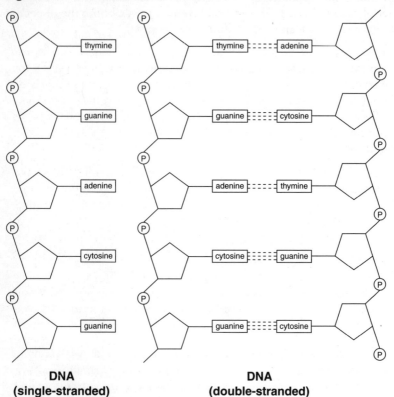

DNA
(single-stranded)

DNA
(double-stranded)

Chemical Reactions in Metabolic Processes

In order for a chemical reaction to take place, the reacting molecules (or atoms) must first collide and then have sufficient energy (activation energy) to trigger the formation of new bonds. Although many reactions can occur spontaneously, the presence of a **catalyst** accelerates the rate of the reaction because it lowers the activation energy required for the reaction to take place. A catalyst is any substance that accelerates a reaction but does not undergo a chemical change itself. Since the catalyst is not changed by the reaction, it can be used over and over again.

Chemical reactions that occur in biological systems are referred to as metabolism. **Metabolism** includes the breakdown of substances (catabolism), the formation of new products (synthesis or anabolism), or the transferring of energy from one substance to another. Metabolic processes have the following characteristics in common:

■ **Enzymes** act as catalysts for metabolic reactions. Enzymes are proteins that are specific for particular reactions. The standard suffix for enzymes is "ase," so it is easy to identify enzymes that use this ending (though some do not). The substance on which the enzyme acts is called the substrate. For example, the enzyme amylase catalyzes the breakdown of the substrate amylose (starch) to produce the product glucose.

The **induced-fit model** describes how enzymes work. Within the protein (the enzyme), there is an active site with which the reactants readily interact because of the shape, polarity, or other characteristics of the active site. The interaction of the reactants (substrate) and the enzyme causes the enzyme to change shape. The new position places the substrate molecules in a position favorable to their reaction and accelerates the formation of the product.

■ **ATP** (adenosine triphosphate) is a common source of activation energy for metabolic reactions. In Figure 1-10, the wavy lines between the last two phosphate groups of the ATP molecule indicate high-energy bonds. When ATP supplies energy to a reaction, it is usually the energy in the last bond that is delivered to the reaction. In the process of giving up this energy, the last phosphate bond is broken and the ATP molecule is converted to ADP (adenosine diphosphate) and a phosphate group (indicated by P_i). In contrast, new ATP molecules are assembled by phosphorylatoin when ADP combines with a phosphate group using energy obtained from some energy-rich molecule (like glucose).

■ **Cofactors** are nonprotein molecules that assist enzymes. A holoenzyme is the union of the cofactor and the enzyme (called an apoenzyme when part of a holoenzyme). If cofactors are organic, they are called coenzymes and usually function to donate or accept some component of a reaction, often electrons. Some vitamins are coenzymes or components of coenzymes. Inorganic cofactors are often metal ions, like Fe^{++}.

Figure 1-10 The high energy bonds of adenosine triphosphate (ATP).

Adenosine Triphosphate (ATP)

Chapter Checkout

Q&A

1. In metabolism, the breakdown of substances is called _____.

2. True or False. The secondary structure of a protein is its three-dimensional shape.

3. Which of the following is true of RNA?

   a. Composed of a nitrogen base, a six-carbon sugar, and a phosphate group
   b. Does not utilize deoxyribose as its sugar
   c. Is often double-stranded
   d. Has thymine, adenosine, cytosine, and uracil as its nucleotides

4. True or False. Phospholipids are composed of a glycerol molecule and three fatty acids.

5. When a substrate binds to an enzyme's active site, this interaction causes the enzyme to change shape. This example of how an enzyme works is called the _____ model.

Answers: 1. catabolism 2. T 3. b 4. F 5. induced-fit

Chapter 2
THE CELL

Chapter Check-In

❑ Discovering the functions and constituents of the plasma membrane

❑ Understanding the various organelles of a cells and their metabolic activities

❑ Identifying the mechanisms by which cells communicate and acquire vital substances

❑ Finding out the differences between mitosis and meiosis

❑ Detailing the process of synthesizing proteins via transcription, RNA processing, and translation

The cell is the smallest functional unit on which all life is built. Therefore, a strong knowledge of the various cellular organelles and their functions is crucial to any physiologist or anatomist. Identifying the components of each cell will not only help to identify the type of cell you're viewing, but also its function.

In order to perform the many diverse metabolic activities in the body, cells need to be able to communicate with one another to regulate growth and development. This is accomplished via genetic instructions, or DNA, contained within each cell. The processes by which this genetic information is replicated and utilized to build proteins are discussed throughout this chapter.

The Cell and Its Membrane

The **cell** is the basic functional unit of all living things. The plasma membrane (cell membrane) bounds the cell and encloses the nucleus and **cytoplasm.** The cytoplasm consist of specialized bodies called organelles suspended in a fluid matrix, the cytosol, which consists of water and dissolved substances such as proteins and nutrients.

The plasma membrane

The **plasma membrane** separates internal metabolic events from the external environment and controls the movement of materials into and out of the cell. The plasma membrane is a double phospholipid membrane (lipid bilayer) with the nonpolar hydrophobic tails pointing toward the inside of the membrane and the polar hydrophilic heads forming the inner and outer faces of the membrane (Figure 2-1).

Proteins and cholesterol molecules are scattered throughout the flexible phospholipid membrane. Proteins may attach loosely to the inner or outer surface of the plasma membrane (peripheral proteins), or they may lie across the membrane, extending from inside to outside (integral proteins). The mosaic nature of scattered proteins within a flexible matrix of phospholipid molecules describes the fluid mosaic model of the cell membrane. Additional features of the plasma membrane follow.

- The phospholipid bilayer is selectively permeable. Only small, uncharged, polar molecules, such as H_2O and CO_2, and hydrophobic molecules—nonpolar molecules like O_2 and lipid soluble molecules such as hydrocarbons—can freely cross the membrane.

Figure 2-1 The phospholipid bilayer of the plasma membrane.

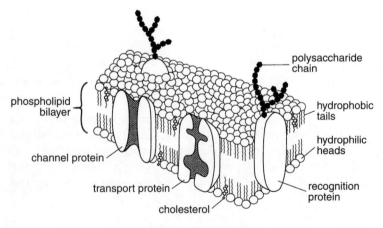

The Plasma Membrane

- **Channel proteins** provide passageways through the membrane for certain hydrophilic (water-soluble) substances such as polar and charged molecules.

■ Transport proteins spend energy (ATP) to transfer materials across the membrane. When energy is used to provide passageway for materials, the process is called active transport.

■ Recognition proteins distinguish the identity of neighboring cells. These proteins have oligosaccharide (short polysaccharide) chains extending out from their cell surface.

■ Adhesion proteins attach cells to neighboring cells or provide anchors for the internal filaments and tubules that give stability to the cell.

■ **Receptor proteins** initiate specific cell responses once hormones or other trigger molecules bind to them.

■ **Electron transfer proteins** are involved in moving electrons from one molecule to another during chemical reactions.

The nucleus and other organelles

Organelles are bodies within the cytoplasm that serve to physically separate the various metabolic activities that occur within cells. They include the following (Figure 2-2):

Figure 2-2 The general organization of a typical cell.

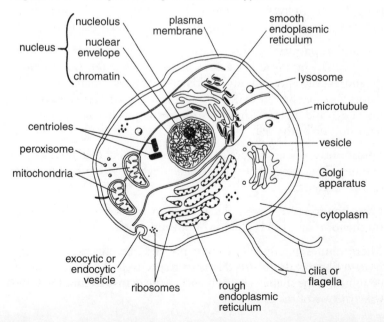

■ The **nucleus** is bounded by the nuclear envelope, a phospholipid bilayer similar to the plasma membrane. The nucleus contains DNA (deoxyribonucleic acid), the hereditary information of the cell. Normally, the DNA is spread out within the nucleus as a threadlike matrix called chromatin. When the cell begins to divide, the chromatin condenses into rod-shaped bodies called chromosomes, each of which, before dividing, is made up of two long DNA molecules and various histone molecules. The histones serve to organize the lengthy DNA, coiling it into bundles called nucleosomes. Also visible within the nucleus are one or more nucleoli, each consisting of DNA in the process of manufacturing the components of ribosomes. Ribosomes are shipped to the cytoplasm where they assemble amino acids into proteins. The nucleus also serves as the site for the separation of chromosomes during cell division.

■ The **endoplasmic reticulum,** or ER, consists of stacks of flattened sacs involved in the production of various materials. In cross-section, they appear as a series of mazelike channels, often closely associated with the nucleus. When ribosomes are present, the ER (called rough ER) attaches polysaccharide groups to polypeptides as they are assembled by the ribosomes. Smooth ER, without ribosomes, is responsible for various activities, including the synthesis of lipids and hormones, especially in cells that produce these substances for export from the cell. In liver cells, smooth ER is involved in the breakdown of toxins, drugs, and toxic byproducts from cellular reactions.

■ A **Golgi apparatus (Golgi complex** or **Golgi body)** is a group of flattened sacs arranged like a stack of bowls. They function to modify and package proteins and lipids into **vesicles,** small, spherically shaped sacs that bud from the ends of a Golgi apparatus. Vesicles often migrate to and merge with the plasma membrane, releasing their contents outside of the cell.

■ **Lysosomes** are vesicles from a Golgi apparatus that contain digestive enzymes. They break down food, cellular debris, and foreign invaders such as bacteria.

■ **Mitochondria** carry out aerobic respiration, a process in which energy (in the form of ATP) is obtained from carbohydrates.

■ **Microtubules,** intermediated filaments, and microfilaments are three protein fibers of decreasing diameter, respectively. All are involved in establishing the shape or movements of the **cytoskeleton,** the internal structure of the cell.

Microtubules are made of the protein tubulin and provide support and mobility for cellular activities. They are found in the spindle apparatus (which guides the movement of chromosomes during cell division) and in flagella and cilia (described later in this list), which project from the plasma membrane to provide motility to the cell.

Intermediate filaments help support the shape of the cell.

Microfilaments are made of the protein actin and are involved in cell motility. They are found in almost every cell, but are predominant in muscle cells and in cells that move by changing shape, such as phagocytes (white blood cells that scour the body for bacteria and other foreign invaders).

■ **Flagella** and **cilia** protrude from the cell membrane and make wavelike movements. Flagella and cilia are classified by their lengths and by their number per cell: Flagella are long and few; cilia are short and many. A single flagellum propels sperm, while the numerous cilia that line the respiratory tract sweep away debris. Structurally, both flagella and cilia consist of microtubules arranged in a "9 + 2" array— that is, nine pairs (doublets) of microtubules arranged in a circle surrounding a pair of microtubules (Figure 2-3).

Figure 2-3   The structural arrangement of various cell specializations.

microtubule doublet

two central microtubules

plasma membrane

Cilium or Flagellum

microtubule triplet

Basal Body

■ **Centrioles** and **basal bodies** act as microtubule organizing centers (MTOCs). A pair of centrioles (enclosed in a centrosome) located outside the nuclear envelope gives rise to the microtubules that make up the spindle apparatus used during cell division. Basal bodies are at the base of each flagellum and cilium and appear to organize their development. Both centrioles and basal bodies are made up of nine triplets arranged in a circle (Figure 2-3).

■ Peroxisomes are organelles common in liver and kidney cells that break down potentially harmful substances. For example, peroxisomes can convert hydrogen peroxide (a toxin made of H_2O_2), to H_2O.

Cell Junctions

The plasma membranes of adjacent cells are usually separated by extracellular fluids that allow transport of nutrients and wastes to and from the bloodstream. In certain tissues, however, the membranes of adjacent cells may join and form a junction. Three kinds of cell junctions are recognized, as follows (Figure 2-4):

Figure 2-4 The three types of cell junctions.

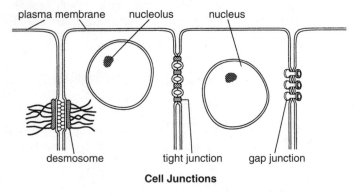

Cell Junctions

■ Desmosomes are protein attachments between adjacent cells. Inside the plasma membrane, a desmosome bears a disk-shaped structure from which protein fibers extend into the cytoplasm. Desmosomes act like spot welds to hold together tissues that undergo considerable stress (such as skin or heart muscle).

■ Tight junctions are tightly stitched seams between cells. The junction completely encircles each cell, preventing the movement of material between the cell. Tight junctions are characteristic of cells lining the digestive tract, where materials are required to pass through cells (rather than intercellular spaces) to penetrate the bloodstream.

■ Gap junctions are narrow tunnels between cells that consist of proteins called connexons. The proteins allow only the passage of ions and small molecules. In this manner, gap junctions allow communication between cells through the exchange of materials or the transmission of electrical impulses.

Movement of Substances

The following terms are used to describe the movement of substances into, out of, and between cells:

■ The movement of substances may occur across a selectively permeable membrane (such as the plasma membrane). A selectively permeable membrane allows only specific substances to pass.

■ The substances, whose movements are being described, may be water (the solvent) or the substance dissolved in the water (the solute).

■ Movement of substances may occur from higher to lower concentrations (down the concentration gradient) or the opposite direction (up or against the gradient).

■ Solute concentrations vary. A solution may be hypertonic (a higher concentration of solutes), hypotonic (a lower concentration of solutes), or isotonic (an equal concentration of solutes) compared to another region.

■ The movement of substances may be passive or active. Usually occurring against the gradient, active movement requires the expenditure of energy.

Passive transport process

Passive transport describes the movement of substances down a concentration gradient and do not require energy consumption.

■ Bulk flow is the collective movement of substances in the same direction in response to a force, such as pressure. Blood moving through a blood vessel is an example of bulk flow.

■ Simple diffusion, or **diffusion,** is the net movement of substances from an area of higher concentration to an area of lower concentration. This movement occurs as a result of the random and constant motion characteristic of all molecules (atoms or ions) and is independent from the motion of other molecules. Since, at any one time, some molecules may be moving against the gradient and some molecules may be moving down the gradient (remember, the motion is random), the word "net" is used to indicate the overall, eventual end-result of the movement. If a concentration gradient exists, then the molecules (which are constantly moving) will eventually become evenly distributed (a state of equilibrium).

- **Osmosis** is the diffusion of water molecules across a selectively permeable membrane. When water moves into a body by osmosis, hydrostatic pressure (osmotic pressure) may build up inside the body.

- **Dialysis** is the diffusion of solutes across a selectively permeable membrane.

- Facilitated diffusion is the diffusion of solutes through channel proteins in the plasma membrane. Note that water can pass freely through the plasma membrane without the aid of specialized proteins.

Active transport processes

Active transport is the movement of solutes against a gradient and requires the expenditure of energy (usually ATP). Active transport is achieved through one of the following two mechanisms:

- Transport proteins in the plasma membrane transfer solute such as small ions (Na^+, K^+, Cl^-, H^+), amino acids, and monosaccharides.

- Vesicles or other bodies in the cytoplasm move macromolecules or large particles across the plasma membrane. Types of vesicular transport include

  Exocytosis, which describes the process of vesicles fusing with the plasma membrane and releasing their contents to the outside of the cell. This process is common when a cell produces substances for export.

  Endocytosis, which describes the capture of a substance outside the cell when the plasma membrane merges to engulf it. The substance subsequently enters the cytoplasm enclosed in a vesicle. There are three kinds of endocytosis:

  Phagocytosis ("cellular eating") occurs when undissolved material enters the cell. The plasma membrane engulfs the solid material, forming a phagocytic vesicle.

Pinocytosis ("cellular drinking") occurs when the plasma membrane folds inward to form a channel allowing dissolved substances to enter the cell. When the channel is closed, the liquid is encircled within a pinocytic vesicle.

Receptor-mediated endocytosis occurs when specific molecules in the fluid surrounding the cell bind to specialized receptors in the plasma membrane. As in pinocytosis, the plasma membrane folds inward and the formation of a vesicle follows. Certain hormones are able to target specific cells by receptor-mediated endocytosis.

Cell Division

Cell division consists of two phases, **nuclear division** followed by **cytokinesis.** Nuclear division divides the genetic material in the nucleus, while cytokinesis divides the cytoplasm. There are two kinds of nuclear division—mitosis and meiosis. Mitosis divides the nucleus so that both daughter cells are genetically identical. In contrast, meiosis is a reduction division, producing daughter cells that contain half the genetic information of the parent cell.

The first step in either mitosis or meiosis begins with the condensation of the genetic material, chromatin, into tightly coiled bodies, the **chromosomes.** Each chromosome is made of two identical halves called sister chromatids, which are joined at the centromere. Each chromatid consists of a single, tightly coiled molecule of DNA. Somatic cells (all body cells except eggs and sperm) are diploid cells because each cell contains two copies of every chromosome. A pair of such chromosomes is called a homologous pair. In a homologous pair of chromosomes, one homologue originates from the maternal parent, the other from the paternal parent. There are 46 chromosomes, 23 homologous pairs, consisting of a total of 92 chromatids.

When a cell is not dividing, the chromatin is enclosed within a clearly defined nuclear envelope, one or more nucleoli are visible within the nucleus, and two centrosomes (each containing two centrioles) lie adjacent to one another outside the nuclear envelope. These features are characteristic of interphase, the nondividing but metabolically active period of the cell cycle (Figure 2-5). When cell division begins, these features change, as described in the following sections.

Figure 2-5 Stages of the cell cycle.

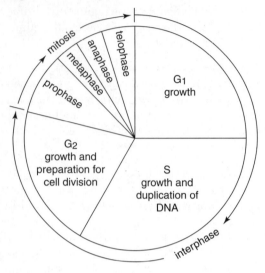

Mitosis

There are four phases in **mitosis** (adjective, mitotic): prophase, metaphase, anaphase, and telophase (Figure 2-6).

■ During prophase, the nucleoli disappear, the chromatin condenses into chromosomes, the nuclear envelope breaks down, and the mitotic spindle is assembled. The development of the mitotic spindle begins as the centrosomes move apart to opposite ends (poles) of the nucleus. As they move apart, microtubules develop from each centrosome, increasing in length by the addition of tubulin units. Microtubules from each centrosome connect to specialized regions in the centromere called kinetochores. Microtubules tug on the kinetochores, moving the chromosomes back and forth, toward one pole, then the other. Within the spindle, there are also microtubules that overlap at the center of the spindle and do not attach to the chromosomes.

■ Metaphase begins when the chromosomes are distributed across the metaphase plate, a plane lying between the two poles of the spindle. Metaphase ends when the microtubules, still attached to the kinetochores, pull each chromosome apart into two chromatids. Each chromatid is complete with a centromere and kinetochores. Once separated from its sister chromatid, each chromatid is called a chromosome. (To count the number of chromosomes at any one time, count the number of centromeres.)

Figure 2-6   The four stages of mitosis.

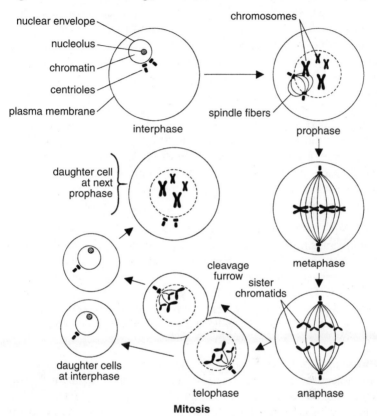

Mitosis

■ Anaphase begins after the chromosomes are separated into chromatids. During anaphase, the microtubules connected to the chromatids (now chromosomes) shorten, effectively pulling the chromosomes to opposite poles. Overlapping microtubules originating from opposite centrosomes but not attached to chromosomes interact to push the poles farther apart. At the end of anaphase, each pole has a complete set of chromosomes, the same number of chromosomes as the original cell. (Since it consists of only one chromatid, each chromosome contains only a single copy of the DNA molecule.)

■ Telophase concludes the nuclear division. During this phase, a nuclear envelope develops around each pole, forming two nuclei. The chromosomes within each of these nuclei disperse into chromatin, and the nuclei reappear. Simultaneously, cytokinesis occurs, dividing the cytoplasm into two cells. Microfilaments form a ring inside the plasma membrane between the two newly forming nuclei. As the microfilaments shorten, they act like purse strings to pull the plasma membrane into the center, dividing the cell into two daughter cells. The groove that forms as the purse strings are tightened is called a cleavage furrow.

Once mitosis is completed and interphase beings, the cell begins a period of growth. Growth begins during the first phase, called G_1, and continues through the S and G_2 phases. Also during the S phase the second DNA molecule for each chromosome is synthesized. As a result of this DNA replication, each chromosome gains a second chromatid. During the G_2 period of growth, materials for the next mitotic division are prepared. The time span from one cell division through G_1, S, and G_2 is called a cell cycle (Figure 2-5).

A cell that begins mitosis in the diploid state—that is, with two copies of every chromosome—will end mitosis with two copies of every chromosome. However, each of these chromosomes will consist of only one chromatid, or one DNA molecule. During interphase, the second DNA molecule is replicated from the first, so that when the next mitotic division begins, each chromosome will, again, consist of two chromatids.

Meiosis

Meiosis (adjective, meiotic) is very similar to mitosis. The major distinction is that meiosis consists of two groups of divisions, meiosis I and meiosis II (Figure 2-7). In meiosis I, homologous chromosomes pair at the metaphase plate and then migrate to opposite poles. In meiosis II, chromosomes spread across the metaphase plate, and sister chromatids separate and migrate to opposite poles. Thus, meiosis II is analogous to mitosis. A summary of each meiotic stage follows.

■ Prophase I begins like prophase of mitosis. The nucleolus disappears, chromatin condenses into chromosomes, the nuclear envelope breaks down, and the spindle apparatus develops. Once the chromosomes are condensed, however, their behavior differs from mitosis. During prophase I, homologous chromosomes pair, a process called synapsis. These pairs of homologous chromosomes are called tetrads (a group of four chromatids) or bivalents. During synapsis, corresponding regions form close associations called chiasmata (singular, chiasma) along nonsister chromatids. Chiasmata are sites where genetic material is exchanged between nonsister homologous chromatids, a process called crossing over. The result contributes to a mixing of genetic material from both parents, a process called genetic recombination.

Figure 2-7 The stages of meiosis.

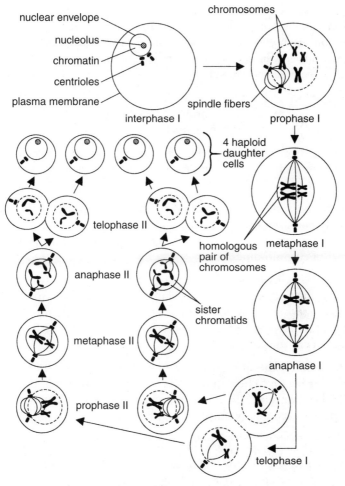

Meiosis

■ At metaphase I, homologous pairs of chromosomes are spread across the metaphase plate. Microtubules extending from one pole are attached to kinetochores of one member of each homologous pair.

Microtubules from the other pole are connected to the second member of each homologous pair.

■ Anaphase I begins when homologues within tetrads uncouple as they are pulled to opposite poles.

■ In telophase I, the chromosomes have reached their respective poles, and a nuclear membrane develops around them. Note that each pole will form a new nucleus that will have half the number of chromosomes, but each chromosome will contain two chromatids. Since daughter nuclei will have half the number of chromosomes, cells that they eventually form will be haploid.

■ Cytokinesis occurs forming two daughter cells. A brief interphase may follow, but no replication of chromosomes occurs. Instead, part II of meiosis begins in both daughter nuclei.

■ In prophase II, the nuclear envelope disappears, and the spindle develops. There are no chiasmata and no crossing over of genetic material as in prophase I.

■ In metaphase II, the chromosomes align singly on the metaphase plate (not in tetrads as in metaphase I). Single alignment of chromosomes is exactly what happens in mitosis—except now there is only half the number of chromosomes.

■ Anaphase II begins as each chromosome is pulled apart into two chromatids by the microtubules of the spindle apparatus. The chromatids (now chromosomes) migrate to their respective poles. Again, this is exactly what happens in mitosis—except now there is only half the number of chromosomes.

■ In telophase II, the nuclear envelope reappears at each pole and cytokinesis occurs. The end result of meiosis is four haploid cells. Each cell contains half the number of chromosomes, and each chromosome consists of only one chromatid.

Meiosis ends with four haploid daughter cells, each with half the number of chromosomes (one chromosome from each homologous pair). These are **gametes**—that is, eggs and sperm. The fusing of an egg and sperm, fertilization (syngamy), gives rise to a diploid cell, the **zygote.** The single-celled zygote then divides by mitosis to produce a multicellular embryo fetus, and after nine months, a newborn infant. Note that one copy of each chromosome pair in the zygote originates from one parent, and the second copy from the other parent. Thus, a pair of homologous chromosomes in the diploid zygote represents both maternal and paternal heritage.

DNA replication

During the S phase of interphase, a second chromatid is assembled. The second chromatid contains the exact same DNA found in the first chromatid. The copying process, called DNA replication, involves separating ("unzipping") the DNA molecule into two strands, each of which serves as a template to assemble a new, complementary strand. The result is two identical double-stranded molecules of DNA that consist of a single strand of old DNA (the template strand) and a single strand of new, replicated DNA (the complementary strand).

Two difficulties must be resolved during replication. First, DNA polymerase, the enzyme that regulates the attachment of new nucleotides to the growing complement strand of DNA, can perform this task in only one direction. The two strands that make up the double helix, however, are oriented in opposite directions (anti-parallel). On one strand, the DNA polymerase can operate continuously toward the replication fork as the DNA double helix is unzipped. The new DNA strand generated in this manner is called the leading complementary strand. Since the nucleotides of the second template strand are oriented in the opposite direction, DNA polymerase must operate away from the replication fork. As a result, replication of the second strand occurs in short segments. As the replication fork advances, the DNA polymerase must repeatedly return to the replication fork to begin replication of short segments of DNA. The enzyme DNA ligase connects these short segments of DNA generating the lagging complementary strand.

The second difficulty that is resolved during replication is that DNA polymerase cannot itself initiate replication. Instead, the enzyme RNA primase initiates each replication using RNA (not DNA) nucleotides. After a short RNA nucleotide segment is generated, DNA polymerase takes over with DNA nucleotides. The initial RNA nucleotides are later replaced with the appropriate DNA nucleotides.

Following are the details of DNA replication. Numbers correspond to events illustrated in Figure 2-8.

1. The enzyme helicase unwinds the DNA helix, producing Y-shaped replication fork.

2. RNA primase initiates DNA replication by producing short segments of RNA nucleotides called RNA primers. (See 2A and 2B in Figure 2-8.)

3. DNA polymerase attaches to the RNA primase and begins adding DNA nucleotides to the complement strand. (In Figure 2-8, the RNA primase is present only on the lagging strand.)

4. The leading complementary strand is assembled continuously as the double-helix of DNA uncoils.

5. The lagging complementary strand is assembled in short segments, which are subsequently joined by DNA ligase. (See 5A, 5B, and 5C in Figure 2-8.)

6. The RNA primers are replaced by DNA nucleotides.

Figure 2-8 The steps of DNA replication.

DNA Replication

Mutations

The replication process of DNA is extremely accurate. However, errors occur when nucleotide bases between DNA strands are occasionally paired incorrectly. In addition, errors in DNA molecules may arise as a result of exposure to radiation (such as ultraviolet or x-ray) or various reactive chemicals. When errors occur, repair mechanisms are available to make corrections.

If a DNA error is not repaired, it becomes a mutation. A mutation is any sequence of nucleotides in a DNA molecule that does not exactly match the original DNA molecule from which it was copied. Mutations include an incorrect nucleotide (substitution), a missing nucleotide (deletion), or an addition nucleotide not present in the original DNA molecule (insertion). When an insertion mutation occurs, it causes all subsequent nucleotides to be displaced one position, producing a frameshift mutation. Radiation or chemicals that cause mutations are called mutagens. Carcinogens are mutagens that activate uncontrolled cell growth (cancer).

Protein synthesis

The DNA in chromosomes contains genetic instructions that regulate development, growth, and the metabolic activities of cells. The DNA instructions determine whether a cell will be that of a pea plant, a human, or some other organism, as well as establish specific characteristics of the cell in that organism. For example, the DNA in a cell may establish that it is a human cell. If during development, it becomes a cell in the iris of any eye, the DNA will direct other information appropriate for its location in the organism, such as the production of brown, blue, or other pigmentation. DNA controls the cell in this manner because it contains codes for polypeptides. Many polypeptides are enzymes that regulate chemical reactions and influence the resulting characteristics of the cell. Thus, from the molecular viewpoint, traits are the end products of metabolic processes regulated by enzymes. A gene is defined as the DNA segment that codes for a particular enzyme or other polypeptide (one-gene-one-polypeptide hypothesis).

The process that describes how enzymes and other proteins are made from DNA is called protein synthesis. There are three steps in protein synthesis—transcription, RNA processing, and translation. In transcription, DNA molecules are used as a template to create RNA. After transcription, RNA processing modifies the RNA molecule with deletions and additions. In translation, the processed RNA molecules are used to assemble amino acids into a polypeptide.

There are three kinds of RNA molecules produced during transcription, as follows.

■ Messenger RNA (mRNA) is a single strand of RNA that provides the template used for sequencing amino acids into a polypeptide. A triplet group of three adjacent nucleotides on the mRNA, called a codon, codes for one specific amino acid. Since there are 64 possible ways that four nucleotides can be arranged in triplet combinations (4 x 4 x 4 = 64), there are 64 possible codons. The genetic code is a table of information that provides "decoding" for each codon—that is, it identifies the amino acid specified by each of the possible 64 codon combinations. For example, the codon composed of the three nucleotides cytosine-guanine-adenine (CGA) codes for the amino acid arginine.

■ Transfer RNA (tRNA) is a short RNA molecule (consisting of about 80 nucleotides) that is used for transporting amino acids to their proper places on the mRNA template. Interactions among various parts of the tRNA molecule result in base-pairings between nucleotides, folding the tRNA in such a way that it forms a three-dimensional molecule. (In two dimensions, a tRNA resembles the three leaflets of a clover leaf.) One end of the tRNA attaches to an amino acid. Another portion of the tRNA, specified by a triplet combination of nucleotides, is the anticodon. During translation, the anticodon of the tRNA base pairs with the codon of the mRNA.

■ Ribosomal RNA (rRNA) molecules are the building blocks of ribosomes. The nucleolus is an assemblage of DNA actively being transcribed into rRNA. Within the nucleolus, various proteins imported from the **cytosol** are assembled with rRNA to form large and small ribosome subunits. Together, the two subunits form a ribosome, which coordinates the activities of the mRNA and tRNA during translation. Ribosomes have three binding sites—one for the mRNA, one for the tRNA that carries a growing polypeptide chain, and one for a second tRNA that delivers the next amino acid that will be inserted into the growing polypeptide chain.

Here are the details of transcription, RNA processing, and protein synthesis (also see Figures 2-9 and 2-10):

■ During transcription, the RNA polymerase attaches to promoter regions on the DNA and beings to unzip the DNA into two strands. (See 1 in Figure 2-9.)

■ As the RNA polymerase unzips the DNA, it assembles new nucleotides using one strand of the DNA as a template. In contrast to the process of DNA replication, the new nucleotides are RNA nucleotides, and only one DNA strand is transcribed . (See 2 in Figure 2-9.)

■ Transcription continues until the RNA polymerase reaches a special sequence of nucleotides that serves as a termination point. The RNA polymerase and the newly created RNA molecule are released. This newly created RNA molecule may be mRNA, tRNA, or rRNA (depending on which DNA segment is transcribed). (See 3 in Figure 2-9.)

Figure 2-9 Transcription and RNA processing.

■ During RNA processing, newly crated mRNA molecules undergo two kinds of alterations. In the first modification, noncoding intervening sequences called introns are removed, leaving only exons, sequences that express a code for a polypeptide. A second modification adds two special sequences—a 5-inch cap to one end of the mRNA and a poly-A tail to the other end. (See 4A, 4B, and 4C in Figure 2-9.)

■ The mRNA, tRNA, and ribosomal subunits are transported across the nuclear envelope and into the cytoplasm. In the cytoplasm, amino acids attach to one end of the tRNAs. (See 5A, 5B, and 5C in Figure 2-10.)

Figure 2-10 The steps involved in protein synthesis.

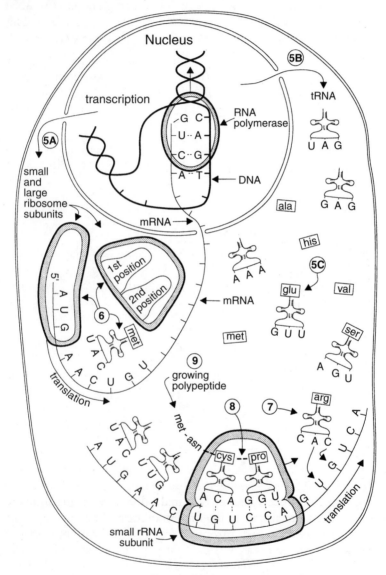

Protein Synthesis

■ Translation begins when the small and large ribosomal subunits attach to one end of the mRNA. Also, a tRNA (with anitcodon UAC) carrying the amino acid methonine attaches to the mRNA (at the "start" codon AUG) within the ribosome. (See 6 in Figure 2-10.)

■ A second tRNA, also bearing an amino acid, arrives and fills a second tRNA position. The codon on the mRNA determines which tRNA (and thus, which amino acid) fills the second position. (Step 7 in Figure 2-10 shows an incoming tRNA approaching a yet to be vacated position.)

■ The amino acid of the first tRNA attaches to the amino acid of the second tRNA, forming a pair of amino acids. Then, the first tRNA is released. The ribosome moves over one codon position, thereby putting the second tRNA in the first position and vacating the second position. (Step 8 in Figure 2-10 shows this process after several tRNAs have delivered amino acids.)

■ A new tRNA (with its amino acid) fills the vacant position. Now, the two amino acids being held by the tRNA in the first position are transferred to the amino acid of the newly arrived tRNA, forming a polypeptide chain of three amino acids. Again, the tRNA in the first position is released, the ribosome moves over one codon position, and the second tRNA position is vacant.

■ The process continues, as new tRNAs bring more amino acids. As each new tRNA arrives, the polypeptide chain is elongated by one new amino acid, growing in sequence and length as dictated by the codons on the mRNA. (See 9 in Figure 2-10.) Eventually, a "stop" codon is encountered, and the ribosome subunits and polypeptide are released.

Once the polypeptide is released, interactions among the amino acids give it its special three-dimensional shape. Subsequent processing by the endoplasmic reticulum or a Golgi body may make final modifications before the protein functions as a structural element or an enzyme.

Chapter Checkout

Q&A

1. _____ provide support and motility to a cell while projecting from the plasma membrane.

2. True or False. Gap junctions allow the transmission of electrical impulses and exchange of material between neighboring cells.

3. In mitosis, which of the following steps is not a characteristic of telophase?

   a. Chromosomes disperse into chromatin.
   b. Microtubules shorten and pull chromosomes to opposite poles.
   c. A cleavage furrow is visible.
   d. The cytoplasm is divided into two cells.

4. True or False. DNA polymerase initiates replication by itself via the utilization of DNA nucleotides.

5. During _____, DNA is unzipped into two separate strands by RNA polymerase.

Critical Thinking Questions

1. Why is genetic recombination so crucial for a species?

2. Take the following single strand of DNA (5' G-C-T-A-C-T 3') throughout the three steps of protein synthesis. What two amino acids would result?

Answers: 1. Microtubules 2. T 3. b 4. F 5. transcription

Chapter 3
TISSUES

Chapter Check-In

❑ Discovering the four basic types of tissue

❑ Listing the general characteristics and functions of the different types of epithelium

❑ Describing the various kinds of connective tissue and their components

❑ Understanding the different kinds of nervous tissue and muscle

Approximately 200 distinctly different types of cells compose the human body. These cells are arranged and organized into four basic tissues that, in turn, are assembled to form organs. When looking at tissue at a microscopic level, the ability to detect the presence and location of the four basic tissues allows identification of the organ at which you are looking. A basic knowledge of the general characteristics and cellular composition of these tissues is essential in **histology,** which is the study of tissues at the microscopic level.

Tissues are groups of similar cells performing a common function. There are four categories of tissues:

■ Epithelial tissue

■ Connective tissue

■ Nervous tissue

■ Muscle tissue

Epithelial Tissue

Epithelial tissue, or epithelium, has the following general characteristics:

- Epithelium consists of closely packed, flat cells. There is little intercellular material.

- The tissue is avascular, or without blood vessels. Nutrient and waste exchange occurs through neighboring connective tissues by diffusion.

- The upper surface of epithelium is free, or exposed to the outside of the body or to an internal body cavity. The basal surface rests on connective tissue. A thin, extracellular layer called the basement membrane forms between the epithelial and connective tissue.

- Cell division in epithelium occurs readily to replace damaged cells.

There are two kinds of epithelial tissues:

- Covering and lining epithelium covers the outside surfaces of the body and lines internal organs.

- Glandular epithelium secretes hormones or other products.

Epithelium that covers or lines

Epithelial tissues that cover or line surfaces are classified by cell shape and by the number of cell layers. The following terms are used to describe these features:

Cell shape:

- Squamous cells are flat. The nucleus, located near the upper surface, gives these cells the appearance of a "fried egg."

- Cuboidal cells are cube or hexagon shaped with a central, circular nucleus. These cells produce secretions (sweat, for example) or absorb substances (digested food, for example).

- Columnar cells are tall with an oval nucleus near the basement membrane. These thick cells serve to protect underlying tissues or may function to absorb substances. Some have microvilli, minute surface extensions, to increase surface area for absorbing substances, while others may have cilia that help move substances over their surface (such as mucus through the respiratory tract).

- Transitional cells range from flat to tall cells that can extend or compress in response to body movement.

Number of cell layers:

■ Simple describes a single layer of cells.

■ Stratified describes epithelium consisting of multiple layers.

■ Pseudostratified describes a single layer of cells of different sizes, giving the appearance of being multilayered.

Names of epithelial tissues include a description of both their shape and their number of cell layers. The presence of cilia may also be identified in their names. For example, simple squamous describes epithelium consisting of a single layer of flat cells. Pseudostratified columnar ciliated epithelium describes a single layer of tall, ciliated cells of more than one size. Stratified epithilieum is named after the shape of the outermost cell layer. Thus, stratified squamous epithelium has outermost layers of squamous cells, even though some inner layers consist of cuboidal or columnar cells. These and other epithelial tissues are illustrated in Figure 3-1.

Glandular epithelium

Glandular epithelium forms two kinds of glands:

■ **Endocrine glands** secrete **hormones** directly into the bloodstream. For example, the thyroid gland secretes the hormone thyroxin into the bloodstream, where it is distributed throughout the body, stimulating an increase in the metabolic rate of body cells.

■ **Exocrine glands** secrete their substances into tubes, or ducts, which carry the secretions to the epithelial surface. Examples of secretions include sweat, saliva, milk, stomach acid, and digestive enzymes.

Exocrine glands are classified according to their structure (see Figure 3-2):

■ Unicellular or multicellular describes a single-celled gland or a gland made of many cells, respectively. A multicellular gland consists of a group of secretory cells and a duct through which the secretions pass as they exit the gland.

■ Branched refers to branching arrangement of secretory cells in the gland.

■ Simple or compound refers to whether the duct of the gland (not the secretory portion) does not branch or branches, respectively.

■ Tubular describes a gland whose secretory cells form a tube, while alvcolar (or acinar) describes secretory cells that form a bulblike sac.

Figure 3-1 Types of epithelial tissues.

stratified squamous

Cells: squamous cells apically,
but basal layers vary from
cuboidal to columnar
Nuclei: centrally located
Functions: protection

cuboidal cells
squamous cells

stratified cuboidal

Cells: two layers
Nuclei: centrally located,
and spherical
Functions: absorption,
secretion

connective tissue
basement membrane

stratified columnar

Cells: single layer of
columnar cells on
several layers of
cuboidal (or many
sided) cells
Nuclei: basal and oval
Functions: protection,
secretion

transitional
epithelium

Cells: vary depending on
stretch, apical cells often
large, round, and bi-
nucleated
Nuclei: centrally located
Functions: distention
(occurs only in
bladder, ureter, and
urethra)

connective tissue
basement membrane

Epithelial Tissues

Exocrine glands are also classified according to their function (see Figure 3-3):

■ In merocrine glands, secretions pass through the cell membranes of
the secretory cells.

■ In apocrine glands, a portion of the cell containing secretions is
released as it separates from the rest of the cell.

■ In holocrine glands, entire secretory cells disintegrate and are released along with their contents.

Figure 3-2 Exocrine glands can be classified as either simple or compound and according to tubular or alveolar structure.

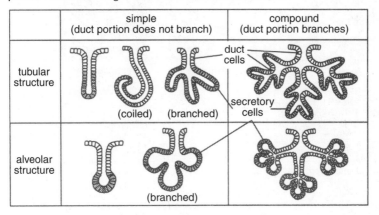

Figure 3-3 Exocrine glands can be classified according to their function.

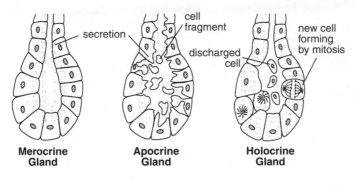

Connective Tissue

A summary of the various kinds of connective tissues is given in Table 3-1 and Figure 3-4. Some general characteristics of connective tissues follow.

Figure 3-4 General characteristics of connective tissues.

Connective Tissues

■ Nerve supply. Most connective tissues have a nerve supply (as does epithelial tissue).

■ Blood supply. There is a wide range of vascularity among connective tissues, although most are well vascularized (unlike epithelial tissues, which are all avascular).

■ Structure. Connective tissue consists of scattered cells immersed in an intercellular material called the matrix. The matrix consists of fibers and ground substance. The kinds and amounts of fiber and ground substance determine the character of the matrix, which in turn defines the kind of connective tissue.

Table 3-1   Kinds of Connective Tissue

Tissue Type	Cells Present	Fibers Present	Matrix Characteristics
Loose Connective Tissue:			
areolar	fibroblasts macrophages adipocytes mast cells plasma cells	collagen elastic reticular	loosely arranged fibers in gelatinous ground substance
adipose	adipocytes	reticular collagen	closely packed cells with a small amount of gelatinous ground substance; stores fat
reticular	reticular cells	reticular	loosely arranged fibers in gelatinous ground substance
Dense Connective Tissue:			
dense regular	fibroblasts	collagen (some elastic)	parallel-arranged bundles of fibers with few cells and little ground substance; great tensile strength

(continued)

Table 3-1 *(continued)*

Tissue Type	Cells Present	Fibers Present	Matrix Characteristics
Dense Connective Tissue:			
dense irregular	fibroblasts	collagen (some elastic)	Irregularly arranged bundles of fibers with few cells and little ground substance; high tensile strength
Cartilage:			
hyaline (gristle)	chondrocytes	collagen (some elastic)	limited ground substance; dense, semisolid matrix
fibrocartilage	chondrocytes	collagen (some elastic)	limited ground intermediate between hyaline cartilage and dense connective tissue
elastic	chondrocytes	elastic	limited ground substance; flexible but firm matrix
Bone (osseous tissue):			
compact (dense)	osteoblasts osteocytes	collagen	rigid, calcified ground substance with osteons (canal systems)
spongy (cancellous)	osteoblasts osteocytes	collagen	rigid, calcified ground substance (no osteons)
Blood & Lymph (vascular tissue):			
blood	erythrocytes leukocytes thrombocytes	"fibers" are soluble proteins that form during clotting	"matrix" is liquid blood plasma

Tissue Type	Cells Present	Fibers Present	Matrix Characteristics
lymph	leukocytes	"fibers" are soluble proteins that form during clotting	"matrix" is liquid blood plasma

■ Cell types. Fundamental cell types, characteristics of each kind of connective tissue, are responsible for producing the matrix. Immature forms of these cells (whose names end in *blast*) secrete the fibers and ground substance of the matrix. Cells that have matured, or differentiated (whose names often end in *cyte*), function mostly to maintain the matrix.

Fibroblasts are common in both loose and dense connective tissues.

Adipocytes, or fat cells, occur in loose connective tissue.

Reticular cells resemble fibroblasts, but have long, cellular processes (extensions). They occur in loose connective tissue.

Chondroblasts and chondrocytes occur in cartilage.

Osteoblasts and osteocytes occur in bone.

Hemocytoblasts occur in the bone marrow and produce erythrocytes (red blood cells), leukocytes (white blood cells), and thrombocytes (blood platelets).

In addition to the fundamental cell types, various leukocytes migrate from the bone marrow to connective tissues and provide various body defense activities.

Macrophages engulf foreign and dead cells.

Mast cells secrete histamine, which stimulates immune responses.

Plasma cells produce antibodies.

■ Fibers. Matrix fibers are protein that provide support for the connective tissue. There are three types:

Collagen fibers, made of the protein collagen, are both tough and flexible.

Elastic fibers, made of the protein elastin, are strong and stretchable.

Reticular fibers, made of thin collagen fibers with a glycoprotein coating, branch frequently to form a netlike (reticulate) pattern.

■ **Ground substance.** Ground substance may be fluid, gel, or solid, and, except for blood, is secreted by the cells of the connective tissue.

Cell adhesion proteins hold the connective tissue together.

Proteoglycans provide the firmness of the ground substance. Hyaluronic sulfate and chondroitin sulfate are two examples.

■ **Classification.** There are five general categories of mature connective tissue:

Loose connective tissue has abundant cells among few or loosely arranged fibers and a sparse to abundant gelatinous ground substance.

Dense connective tissue has few cells among a dense network of fibers with little ground substance.

Cartilage has cells distributed among fibers in a firm jellylike ground substance. Cartilage is tough, but flexible, avascular, and without nerves.

Bone has cells distributed among abundant fibers in a solid ground substance containing minerals, mostly calcium phosphate. Bone is organized in units, called osteons (Haversian system). Each osteon consists of a central canal (Haversian canal), which contains blood vessels and nerves, surrounded by concentric rings (lamellae) of hard matrix and collagen fibers. Between the lamellae are cavities (lacunae) that contain bone cells (osteocytes). Canals (canaliculi) radiate from the central canal and allow nutrient and waste exchange with the osteocytes.

Blood is composed of various blood cells and cell fragments (platelets) distributed in a fluid matrix called blood plasma.

■ **Tissue origin.** All mature connective tissues originate from embryonic connective tissue. There are two kinds of embryonic connective tissues:

Mesenchyme is the origin of all mature connective tissues.

Mucous connective tissue is a temporary tissue formed during embryonic development.

An **epithelial membrane** is a combination of epithelial and connective tissues working together to perform a specific function. As such, it acts as an organ. There are four principle types of epithelial membranes:

- Serous membranes line interior organs and cavities. The serous membranes that line the heart, lungs, and abdominal cavities and organs are called the pericardium, pleura, and peritoneum, respectively.

- Mucous membranes line body cavities that open to the outside of the body. These include the nasal cavity and the digestive, respiratory, and urogenital tracts.

- Synovial membranes line the cavities at bone joints.

- The cutaneous membrane is the skin.

Nervous Tissue

Nervous tissue consists of two kinds of nerve cells:

- Neurons are the basic structural unit of the nervous system. Each cell consists of the following parts (see Figure 3-5).

  The cell body contains the nucleus and other cellular organelles.

  The dendrites are typically short, slender extensions of the cell body that receive stimuli.

  The axon is typically a long, slender extension of the cell body that sends stimuli.

- Neuroglia, or glial cells, provide support functions for the neurons, such as insulation or anchoring neurons to blood vessels.

Figure 3-5 A neuron is a basic structural unit of the nervous system containing a cell body, dendrites, and an axon.

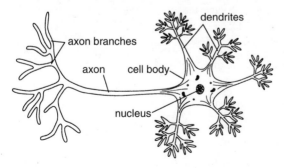

Muscle Tissue

There are three kinds of **muscle tissues** (see Figure 3-6):

- **Skeletal muscle** consists of long cylindrical cells that, under a microscope, appear striated with bands perpendicular to the length of the cell. The many nuclei in each cell (multinucleated cells) are located near the outside along the plasma membrane. Skeletal muscle is attached to bones and causes movements of the body. Because it is under your conscious control, it is also called voluntary muscle.

- **Cardiac muscle,** like skeletal muscle, is striated. However, cardiac muscle cells have single, centrally located nucleus, and the muscle fibers branch often. Where two cardiac muscle cells meet, they form an intercalated disc containing gap junctions, which bridge the two cells.

- **Smooth muscle** consists of cells with a single, centrally located nucleus. The cells are elongated with tapered ends and do not appear striated. Smooth muscle lines the walls of blood vessels and certain organs such as the digestive and urogenital tracts, where it serves to advance the movement of substances. Smooth muscle is called involuntary muscle because it is not under direct conscious control.

Figure 3-6 Three kinds of muscle tissue exist: skeletal muscle, cardiac muscle, and smooth muscle.

Chapter Checkout

Q&A

1. _____ are a type of epithelial specialization that greatly increases surface area to enhance absorption.

2. Which of the following statements is true concerning skeletal muscle?
   a. have centrally located nuclei
   b. form an intercalated disk between two cells
   c. line the walls of the digestive tract
   d. contain multinucleated cells

3. True or False. Blood is a type of connective tissue.

Critical Thinking Questions

1. What type of epithelium, connective tissue, and muscle would you expect to see in the duodenum?

2. Identify three locations in the body where the mechanical protection of stratified squamous epithelium is needed and found.

Answers: 1. Microvilli 2. d 3. T

Chapter 4

THE INTEGUMENTARY SYSTEM

Chapter Check-In

❑ Identifying the individual layers of each component of the skin

❑ Distinguishing the variety of accessory organs that reside within the skin

❑ Comprehending the physiological functions of the skin

The skin is far more than just the outer covering of human beings; it is an organ just like the heart, lung, or liver. Besides providing a layer of protection from pathogens, physical abrasions, and radiation from the sun, the skin serves many functions as well as playing a vital role in homeostasis. This includes maintaining a constant body temperature via the acts of sweating or shivering and by making you aware of external stimuli through information perceived within the touch receptors located within the integumentary system. It only takes one visit to a burn unit to see the value of skin and the many complications that arise when this organ is compromised. In this chapter, you see the diverse actions of the integumentary system, as well as the composition of the skin that allows it to perform these many functions.

The Skin and Its Functions

The skin, or integument, is considered an organ because it consists of two tissues: epithelial and connective. In addition, accessory organs, such as glands, hair, and nails, are present, and together with the skin make up the **integumentary system.** A section of skin with various accessory organs is shown in Figure 4-1.

The skin consists of two layers, the **epidermis** and the underlying **dermis.** Although technically not part of the skin, the **hypodermis** (subcutaneous layer, or superficial fascia) lies beneath the dermis.

Figure 4-1   A section of skin with various accessory organs.

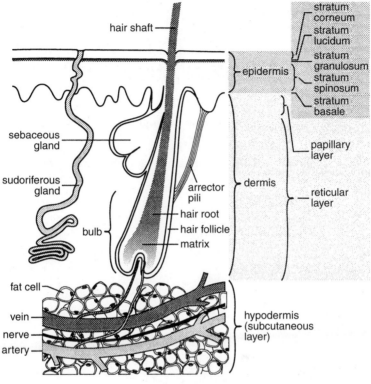

Skin and Hypodermis

The skin performs a variety of functions:

■ Protection is provided against biological invasion, physical damage, and ultraviolet radiation.

■ Sensation for touch, pain, and heat is provided by nerve endings.

■ Thermoregulation is supported through the sweating and regulation of blood flow through the skin.

■ Metabolism of vitamin D occurs in the skin.

■ Storage of blood that can be shunted to other parts of the body when needed takes place in the skin.

■ Excretion of salts and small amounts of wastes (ammonia and urea) occurs with the production of sweat.

The Epidermis

The epidermis consists of stratified squamous epithelium. Four cell types are present:

- Keratinocytes produce keratin, a protein that hardens and waterproofs the skin. Mature keratinocytes at the skin surface are dead and filled almost entirely with keratin.

- Melanocytes produce melanin, a pigment that protects cells from ultraviolet radiation. Melanin from the melanocytes is transferred to the keratinocytes.

- Langerhans cells are phagocytic macrophages that interact with white blood cells during an immune response.

- Merkel cells occur deep in the epidermis at the epidermal-dermal boundary. They form Merkel discs, which, in association with nerve endings, serve a sensory function.

Five layers make up the epidermis:

- The stratum corneum contains many layers of dead, anucleate keratinocytes completely filled with keratin. The outermost layers are constantly shed.

- The stratum lucidum contains two to three layers of anucleate cells. This layer is usually apparent only in thick skin (palms of hands and soles of feet).

- The stratum granulosum contains two to four layers of cells held together by desmosomes. These cells contain keratohyaline granules, which contribute to the formation of keratin in the upper layers of the epidermis.

- The stratum spinosum contains eight to ten layers of cells connected by desmosomes. These cells are moderately active in mitosis.

- The stratum basale contains a single layer of columnar cells actively dividing by mitosis to produce cells that migrate into the upper epidermal layers and ultimately to the surface of the skin.

The Dermis

The second layer of the skin, the dermis, consists of various connective tissues. As connective tissue, it contains fibroblasts and macrophages within a gelatinous matrix containing collagen, elastic, and reticular fibers. The structure provides strength, extensibility (the ability to be stretched), and elasticity (the ability to return to its original form).

The dermis consists of two layers:

- The papillary layer is a thin, outer layer with fingerlike projections called dermal papillae that protrude into the epidermis. In the hands and feet, the dermal papillae generate epidermal ridges (sweat from the epidermal ridges leaves fingerprints).

- The reticular layer is a thick layer, below the papillary layer, that makes up most of the dermis.

The Hypodermis

The hypodermis (subcutaneous layer, or superficial fascia) lies between the dermis and underlying tissues and organs. It consists of mostly adipose tissue and is the storage site of most body fat. It serves to fasten the skin to the underlying surface, provides thermal insulation, and absorbs shocks from impacts to the skin.

Accessory Organs of the Skin

The following accessory organs (skin derivatives) are embedded in the skin:

- Hairs are elongated filaments of keratinized epithelial cells that arise and emerge from the skin of mammals. Hair is composed of the following structures:

  The hair shaft is the portion of the hair that is visible on the surface of the skin.

  The hair root is the portion of the hair that penetrates the skin (epidermis and dermis).

  The hair follicle is the sheath that surrounds the hair in the skin.

  The bulb is the base of the hair follicle.

  The matrix is the bottom of the hair follicle (located within the bulb). Here, cells are actively dividing, producing new hair cells. As these cells differentiate, they produce keratin and absorb melanin from nearby melanocytes. As younger cells are produced below them, the more mature cells are pushed upward, where they eventually die. The keratin they leave behind contributes to the growth of the hair. The color of the hair is determined by the pigments absorbed from the melanocytes.

The arrector pili is a smooth muscle that is attached to the hair follicle. When the muscle contracts, the hair becomes erect; in humans, "goose bumps" are produced.

■ Nails are keratinized epithelial cells. The semilunar lighter region of the nail, the lunula, is the area of new nail growth. Below the lunula, the nail matrix is actively producing nail cells, which contribute to the growth of the nail.

■ Sudoriferous (sweat) glands secrete sweat. Sweat consists of water with various salts and other substances. There are four kinds of sudoriferous glands:

Eccrine glands occur under most skin surfaces and secrete a watery solution through pores (openings at the skin surface), which serve to cool the skin as it evaporates.

Apocrine glands occur under skin surfaces of the armpits and pubic regions and, beginning with puberty, secrete a solution in response to stress or sexual excitement. The solution, more viscous than that secreted by eccrine glands, is secreted into hair follicles.

Ceruminous glands secrete cerumen (earwax) into the external ear canal. Wax helps to impede the entrance of foreign bodies.

Mammary glands produce milk that is secreted through the nipples of the breasts.

■ Sebaceous (oil) glands secrete sebum, an oily substance, into hair follicles or sometimes through skin surface pores. Sebum inhibits bacterial growth and helps prevent drying of hair and skin. An accumulation of sebum in the duct of a sebaceous gland produces whiteheads, blackheads (if the sebum oxidizes), and acne (if the sebum becomes infected by bacteria).

Chapter Checkout

Q&A

1. Skin is very important in the metabolism of what important nutrient?
2. True or False. The dermis is the storage site of most body fat and plays an important role in thermal regulation.
3. Which type of epithelium is present within skin?
    a. transitional epithelium
    b. stratified cuboidal
    c. stratified squamous
    d. pseudostratified columnar
    e. simple squamous

Answers: 1. vitamin D 2. F 3. c

Chapter 5

BONES AND SKELETAL TISSUES

Chapter Check-In

❑ Comprehending the various functions and types of bone

❑ Understanding the difference between compact and spongy bone

❑ Listing the main features of long, short, irregular, and flat bones

❑ Appreciating the basics of how bone develops, grows, and remodels

Bone is often stereotyped as simply a protective and supportive framework for the body. Though it does perform these functions, bone is actually a very dynamic organ that is constantly remodeling and changing shape to adapt to the daily forces placed upon it. Moreover, bone stores crucial nutrients, minerals, and lipids and produces blood cells that nourish the body and play a vital role in protecting the body against infection. All these functions make approximately 206 bones of the human body an organ that is essential to your daily existence.

Functions of Bones

The **skeletal system** consists of bones, cartilage, and the membranes that line the bones. Each bone is an organ that includes connective tissue (bone, blood, cartilage, adipose tissue, and fibrous connective tissue), nervous tissue, and muscle and epithelial tissues (within the blood vessels).

Bones have many functions, including

■ *Support.* Bones provide a framework for the attachment of muscles and other tissues.

■ *Protection.* Bones such as the skull and rib cage protect internal organs from injury.

- *Movement.* Bones enable body movements by acting as levers and points of attachment for muscles.

- *Mineral storage.* Bones serve as a reservoir for calcium and phosphorus, essential minerals for various cellular activities throughout the body.

- *Blood cell production.* The production of blood cells, or **hematopoiesis,** occurs in the red marrow found within the cavities of certain bones.

- *Energy storage.* Lipids (fats) stored in adipose cells of the yellow marrow serve as an energy reservoir.

Types of Bones

Bones come in several different types. Long bones are longer than they are wide. The length of the bone, or shaft, widens at the extremities (ends). Short bones are cubelike, about as long as they are wide. Flat bones, such as ribs or skull bones, are thin or flattened. Irregular bones, such as vertebrae, facial bones, and hip bones, have specific shapes unlike the other types of bones.

The following two bone types are usually classified separately:

- Sesamoid, or round bones, such as the kneecap, are found embedded within certain tendons.

- Sutural, or Wormian bones, occur between the sutures (joints) of the cranial bones of the skull.

Bone Structure

There are two kinds of bone tissue (see Figure 5-1):

- **Compact bone** is the hard material that makes up the shaft of long bones and the outside surfaces of other bones. Compact bone consists of cylindrical units called osteons (Haversian systems). Each osteon contains concentric lamellae (layers) of hard, calcified matrix with osteocytes (bone cells) lodged in lacunae (spaces) between the lamellae. Smaller canals, or canaliculi, radiate outward from a central canal (Haversian canal), which contains blood vessels and nerve fibers. Osteocytes within an osteon are connected to each other and to the central canal by fine cellular extensions. Through these cellular extensions, nutrients and wastes are exchanged between the osteocytes and the blood vessels. Perforating canals (Volkmann's canals) provide channels that allow the blood vessels that run through the central canals to connect to the blood vessels in the periosteum that surrounds the bone.

Figure 5-1 Main features of a long bone.

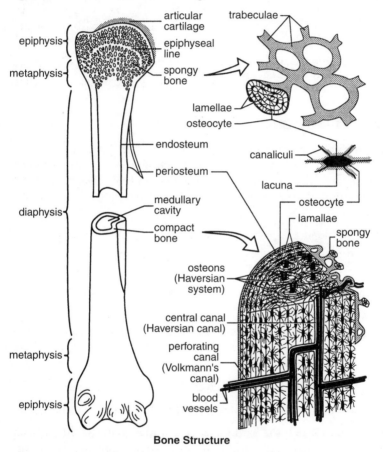

Bone Structure

■ **Spongy bone** consists of thin, irregularly shaped plates called trabeculae, arranged in a latticework network. Trabeculae are similar to osteons in that both have osteocytes in lacunae that lie between calcified lamellae. As in osteons, canaliculi present in trabeculae provide connections between osteocytes. However, since each trabecula is only a few cell layers think, each osteocyte is able to exchange nutrients with nearby blood vessels. Thus, no central canal is necessary.

Here are the main features of a long bone (refer to Figure 5-1):

- The diaphysis, or shaft, is the long tubular portion of long bones. It is composed of compact bone tissue.

- The epiphysis (plural, epiphyses) is the expanded end of a long bone.

- The metaphysis is the area where the diaphysis meets the epiphysis. It includes the epiphyseal line, a remnant of cartilage from growing bones.

- The medullary cavity, or marrow cavity, is the open area within the diaphysis. The adipose tissue inside the cavity stores lipids and forms the yellow marrow.

- Articular cartilage covers the epiphysis where joints occur.

- The periosteum is the membrane covering the outside of the diaphysis (and epiphyses where articular cartilage is absent). It contains osteoblasts (bone-forming cells), osteoclasts (bone-destroying cells), nerve fibers, and blood and lymphatic vessels. Ligaments and tendons attach to the periosteum.

- The endosteum is the membrane that lines the marrow cavity.

Here are the main features of short, flat, and irregular bones:

- In short and irregular bones, spongy bone tissue is encircled by a thin layer of compact bone tissue.

- In flat bones, the spongy bone tissue is sandwiched between two layers of compact bone tissue. The spongy bone tissue is called the diploë.

- Periosteum covers the outside layer of compact bone tissue.

- Endosteum covers the trabeculae that fill the inside of the bone.

- In certain bones (ribs, vertebrae, hip bones, sternum), the spaces between the trabeculae contain red marrow, which is active in hematopoiesis.

Bone Development

The skeleton arises from fibrous membranes and hyaline cartilage during the first month of embryonic development. These tissues are replaced with bone by two different bone-building, or ossification, processes.

The first process, called intramembranous ossification, occurs when fibrous membranes are replaced by bone tissue. The process, occurring only in certain flat bones, is summarized in two basic steps:

■ Spongy bone tissue begins to develop at sites within the membranes called centers of ossification.

■ Red bone marrow forms within the spongy bone tissue, followed by the formation of compact bone on the outside.

The second ossification process, called endochondral ossification, occurs when hyaline cartilage is replaced by bone tissue. The process, occurring in most bones of the body, follows these steps:

■ At a primary ossification center, in the center of a cartilage model, hyaline cartilage breaks down, forming a cavity.

■ A periosteal bud, consisting of osteoblasts, osteoclasts, red marrow, nerves, and blood and lymph vessels, invades the cavity. The osteoblasts produce spongy bone tissue.

■ A medullary cavity forms as osteoclasts break down the newly produced spongy bone tissue. The medullary cavity expands as it follows the spread of the primary ossification center to the ends of the bone.

■ Compact bone tissue replaces cartilage on the outside of the bone.

■ In long bones, secondary ossification centers form in the epiphyses. As in the shaft, a periosteal bud develops. However, the spongy bone tissue that subsequently develops is not replaced by a medullary cavity.

■ Articular cartilage is formed from cartilage remaining on the outside of the epiphyses.

■ The epiphyseal plate is formed from cartilage remaining between the expanding primary and secondary ossification centers.

Bone Growth

Bones elongate as chondrocytes in the cartilage of the epiphyseal plate divide. These cell divisions produce new cartilage within the epiphyseal plate bordering the epiphyses. At the other end of the epiphyseal plate, bordering the diaphysis, older cartilage is broken down by invading osteoclasts and is eventually replaced by the expanding medullary cavity.

Bone Homeostasis

Remodeling is the process of creating new bone and removing old bone. It occurs constantly in growing children as well as in adults in the following situations:

- Remodeling occurs when bones grow, redistributing bone tissue to maintain the shape and structure of the bone.

- Remodeling occurs in response to new stresses applied to a bone, increasing bone strength by adding new bone tissue where appropriate.

- Remodeling occurs when calcium stored in the bones is removed for metabolic processes in other parts of the body. Similarly, remodeling occurs when excess calcium is returned to the bone reservoir.

- Remodeling occurs during the repair of broken bones.

Surface Features of Bones

The surfaces of bones bear projections, depressions, ridges, and various other features. A process (projection) on one bone may fit with a depression on a second bone to form a joint. Another process allows for the attachment of a muscle or ligament. Grooves and openings provide passageways for blood vessels or nerves. A list of the various processes and other surface features appears in Table 5-1.

Table 5-1   Various Processes & Other Surface Features

General	Process	Projection or Prominence on a Bone
Processes that help form joints	Condyle	Large, rounded articular process
	Facet	Smooth, flat surface
	Head	Enlarge portion at an end of a bone
	Ramus	Branch or extension of a bone

(continued)

Table 5-1 *(continued)*

General	Process	Projection or Prominence on a Bone
Processes that provide for the attachment of muscles and ligaments	Crest	Narrow ridge
	Epicondyle	Process on or above a condyle
	Linea (line)	Narrow ridge (less prominent than a crest)
	Spine	Sharp or pointed process (spinous process)
	Trochanter	Large, irregularly shaped process (found only on the femur)
	Tubercle	Small, knoblike process
	Tuberosity	Large, knoblike process
Depressions or openings (may provide passageways for blood vessels and nerves)	Fissure	Narrow opening
	Foramen	Round opening
	Fossa	Shallow depression
	Fovea	Pitlike depression
	Fontanel	Membrane-covered spaces between skull bones
	Meatus	Tubelike passage
	Sinus	Interior cavity
	Sulcus (or groove)	Long, narrow depression

Chapter Checkout

Q&A

1. In spongy bone, there is a meshwork of thin, irregularly shaped plates called _____.

2. True or False. Some bones are completely embedded in tendons.

3. Which of the following bone (or bones) represent(s) a flat bone?
   a. the frontal bone
   b. the humerus
   c. the ninth rib
   d. a cervical vertebra

4. True or False. The skeleton arises from fibrous membranes and elastic cartilage during the early stages of embryonic development.

5. The _____ is a remnant of cartilage found in growing long bones

Answers: 1. trabeculae 2. T (sesamoid bones) 3. a & c 4. F 5. epiphyseal line

Chapter 6

THE SKELETAL SYSTEM

Chapter Check-In

❑ Understanding the differences between the axial and appendicular skeleton

❑ Identifying the bony composition of each part of the skeleton

❑ Listing the names of the various bones of the body

Knowing the skeletal organization of each part of the body is crucial in anatomy. Though memorizing the names and locations of all 206 bones is quite a tedious process, this knowledge can be very valuable. This is especially true if you are planning a future in science, want to understand the language your physician is using, hope to do well in an anatomy course, or peruse scientific literature.

Organization of the Skeleton

The bones of the body are categorized into two groups, the **axial skeleton** and the **appendicular skeleton.** The bones of the axial skeleton revolve around the vertical axis of the skeleton, while the bones of the appendicular skeleton make up the limbs that have been appended to the axial skeleton. A list of the bones in the axial and appendicular skeletons is given in Tables 6-1 and 6-2, respectively. Major bones of both skeletons are shown in Figure 6-1.

Table 6-1   Axial Skeleton Bones (80 Bones)

General Description	Name of Bone	No. of Bones	Additional Information
cranium (8)	frontal	1	
	parietal	2	
	temporal	2	
	sphenoid	1	
	ethmoid	1	
	occipital	1	
facial bones (14)	mandible	1	lower jawbone
	maxilla	2	upper jawbone
	zygomatic bone	2	cheek bones
	nasal bone	2	
	lacrimal bone	2	
	palatine	2	
	inferior nasal concha	2	
	vomer	1	
hyoid	hyoid	1	
ear ossicles	malleus, incus, stapes	6	2 each
vertebral column (26)	cervical vertebrae	7	C_1-C_7
	thoracic vertebrae	12	T_1 - T_{12}
	lumbar vertebrae	5	L_1 - L_5
	sacrum	1	5 fused S_1 - S_5
	coccyx	1	4 fused
thorax (thoracic cage or bony thorax)	sternum	1	
	true ribs (7 pair)	14	vertebrosternal
	false ribs (3 pair)	6	vertebrochondral
	false ribs (floating ribs)	4	vertebral ribs

Table 6-2   Appendicular Skeleton Bones (126 Bones)

General Description	Name of Bone	No. of Bones	Additional Information
pectoral girdle (shoulder girdle)	clavicle	2	collarbone
	scapula	2	shoulder blade
upper limb (60)	*humerus*	*2*	*1 per arm*
	ulna	*2*	*1 per forearm*
	radius	*2*	
	carpals (16)		*8 per wrist*
	1. scaphoid	*2*	
	2. trapezium	*2*	
	3. capitate	*2*	
	4. trapezoid	*2*	
	5. lunate	*2*	

General Description	Name of Bone	No. of Bones	Additional Information
	6. triquetrum	2	
	7. pisiform	2	
	8. hamate	2	
	metacarpals	10	5 in each arm
	phalanges	28	3 per digit (pollex has 2)
pelvic girdle (hip) (2)	coxal bones (os coxae) 1. ilium 2. ischium 3. pubis	2	hip bones (3 fused pairs)
lower limb (60)	femur	2	
	patella	2	
	tibia	2	
	fibula	2	
lower limb (60) 1. talus 2. calcaneus 3. cuboid 4. navicular 5. medial cuneiform 6. interm. cuneiform 7. lateral cuneiform	tarsal (14) 2 2 2 2 2 2 2		
lower limb (60)	metatarsals	2	5 per foot
	phalanges	2	3 per toe (hallux has 2)

Skull: Cranium and Facial Bones

The skull consists of 8 cranial bones and 14 facial bones. The bones are listed in Table 6-1, but note that only six types of cranial bones and eight types of facial bones are listed because some of the bones (as indicated in the table) exist as pairs.

The bones of the skull provide protection for the brain and the organs of vision, taste, hearing, equilibrium, and smell. The bones also provide attachment for muscles that move the head and control facial expressions and chewing.

Figures 6-2 and 6-3 illustrate specific characteristics of these bones, while some general features of the skull follow:

■ Sutures are immovable interlocking joints that join skull bones together.

■ Fontanels are spaces between cranial bones that are filled with fibrous membranes. The spaces provide pliability for the skull when it passes through the birth canal and for brain growth during infancy. Bone growth eventually fills the spaces by age two.

■ Sutural (Wormian) bones are very small bones that develop within sutures. Their number and location vary.

■ The cranial vault denotes the top, sides, front, and back of the cranium. The cranial floor (base) denotes the bottom of the cranium.

■ Cranial fossae are three depressions in the floor of the cranium. These fossae, called the anterior, middle, and posterior cranial fossae, provide spaces that accommodate the shape of the brain.

■ The nasal cavity is formed by cartilage and several bones. Air entering the cavity is warmed and cleansed by mucus lining the cavity.

■ Sinuses (paranasal sinuses) are mucus-lined cavities inside cranial and facial bones that surround the nasal cavity. The cavities secrete mucus that drains into the nasal cavity. The cavities also act as resonance chambers that enhance vocal (and singing) quality.

Figure 6-1 Major bones of the axial and appendicular skeletons.

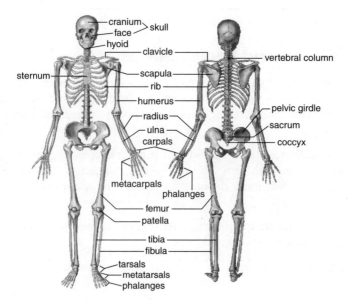

Figure 6-2 The right lateral view and anterior view of the skull's bones.

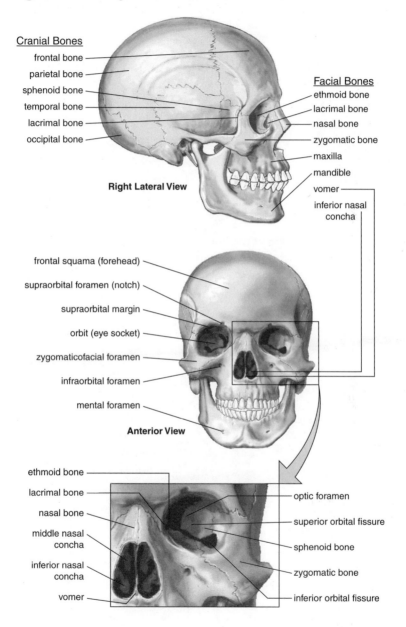

Cranial Bones
frontal bone
parietal bone
sphenoid bone
temporal bone
lacrimal bone
occipital bone

Right Lateral View

Facial Bones
ethmoid bone
lacrimal bone
nasal bone
zygomatic bone
maxilla
mandible
vomer
inferior nasal concha

frontal squama (forehead)
supraorbital foramen (notch)
supraorbital margin
orbit (eye socket)
zygomaticofacial foramen
infraorbital foramen
mental foramen

Anterior View

ethmoid bone
lacrimal bone
nasal bone
middle nasal concha
inferior nasal concha
vomer

optic foramen
superior orbital fissure
sphenoid bone
zygomatic bone
inferior orbital fissure

Figure 6-3 The sagittal section and interior view of the skull's bones.

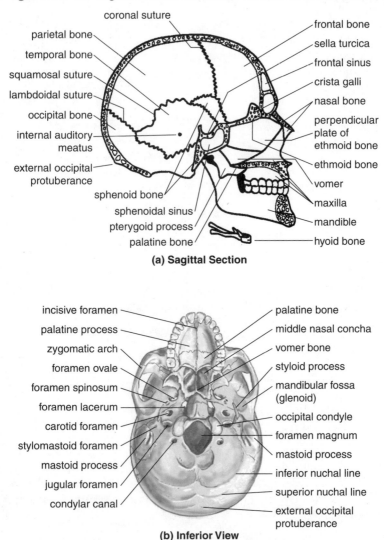

coronal suture

parietal bone
temporal bone
squamosal suture
lambdoidal suture
occipital bone
internal auditory meatus
external occipital protuberance
sphenoid bone
sphenoidal sinus
pterygoid process
palatine bone

frontal bone
sella turcica
frontal sinus
crista galli
nasal bone
perpendicular plate of ethmoid bone
ethmoid bone
vomer
maxilla
mandible
hyoid bone

(a) Sagittal Section

incisive foramen
palatine process
zygomatic arch
foramen ovale
foramen spinosum
foramen lacerum
carotid foramen
stylomastoid foramen
mastoid process
jugular foramen
condylar canal

palatine bone
middle nasal concha
vomer bone
styloid process
mandibular fossa (glenoid)
occipital condyle
foramen magnum
mastoid process
inferior nuchal line
superior nuchal line
external occipital protuberance

(b) Inferior View

Hyoid Bone

Located in the neck, the hyoid bone is isolated from all other bones (refer to Figure 6-3). It is connected by ligaments to the styloid processes of the temporal bones. Muscles from the tongue, neck, pharynx, and larynx that attach to the hyoid bone contribute to the movements involved in swallowing and speech.

Vertebral Column

The vertebral column (spine) consists of 26 vertebrae bones (Table 6-3). It provides support for the head and trunk of the body, protection for the spinal cord, and connecting points for the ribs and muscles.

A typical vertebra has the following characteristics (see Figure 6-4).

- The body (centrum) is the disc-shaped, anterior portion that gives strength to the bone.

- The vertebral arch is a bony ring behind the vertebral body. The opening in the ring is the vertebral foramen, the passageway for the spinal cord. The vertebral canal is the continuous passageway formed by the vertebral foramina of successive vertebrae.

- The pedicles and the laminae form the anterior and posterior sides, respectively, of the vertebral arch.

- Seven processes project from the vertebral arch:

  A spinous process projects posteriorly from the vertebral arch. Muscle and ligaments attach here.

  Two transverse processes project from the vertebral arch, one from each side, at each of the junctures of the pedicles and laminae. Muscles and ligaments attach here. Each transverse process of cervical vertebrae contains a transverse foramen through which blood vessels pass to the brain.

  Two superior articular processes project from the superior surface of the vertebral arch, one from each of the pediclelamina junctions. These processes articulate (form joints) with the preceding vertebrae.

  Two inferior articular processes project from the inferior surface of the vertebral arch, one from each of the pediclelamina junctions. These processes articulate (form joints) with the following vertebra.

- The intervertebral foramina are notches in the superior and inferior surfaces of each pedicle of the vertebral arch. Adjacent notches of successive vertebrae form a passage for nerves that leave the spinal cord and emerge outside the vertebral column.

- Intervertebral discs separate adjacent vertebrae. Each disc consists of an outer ring of fibrocartilage (annulus fibrosus) surrounding a semifluid cushion (nucleus pulposus) that provides elasticity and compressibility.

The vertebral column is divided into four regions, each region contributing to the alternating concave and convex curves of the spine (see Figure 6-4). As the vertebrae progress down the column, their bodies get more massive, enabling them to bear more weight.

The sacrum is a triangular bone below the last lumbar vertebra (see Figure 6-4). It is formed by the fusion of five vertebrae (S_1-S_5).

The coccyx, formed by four fused vertebrae, is a small triangle-shaped bone that attaches to the bottom of the sacrum (see Figure 6-4).

Table 6-3   Bones of the Vertebral Column

Vertebra	Body	Spinous Process	Transverse Process	Vertebral Process
Cervical				
C_1 (atlas)	none, bony ring	none	with transverse foramen	large; lightbulb-shaped
C_2 (axis)	relatively small; with dens (odontoid process)	bifid	with transverse foramen	large; heart-shaped
C_3-C_6	relatively small; oval	bifid	with transverse foramen	large; triangular

Vertebra	Body	Spinous Process	Transverse Process	Vertebral Process
Cervical				
C₇ (vertebra prominens)	relatively small; oval	prom- inently long; not bifid	with transverse foramen	large; triangular
Thoracic				
T₁ - T₁₀	larger than C₂-C₇, heart- shaped; 2 facets or demifacets for articulating with rib head	long; points down	with facets for articulating with rib tubercle	circular
T₁₁-T₁₂	larger than C₂-C₇, heart- shaped; 1 demifacets for articulating with rib head	long; points down	no facets for rib joints	circular
Lumbar				
L₁-L₅	largest of all vertebrae; kidney- shaped	short, thick; points horizontally		
Sacrum				
5 fused S₁-S₅	fusion of 5 vertebrae forms a triangular bone	fuse to become the median sacral crest	becomes the lateral sacral crest	becomes sacral canal
Coccyx				
4 fused				

Figure 6–4 The four regions of the vertebral column.

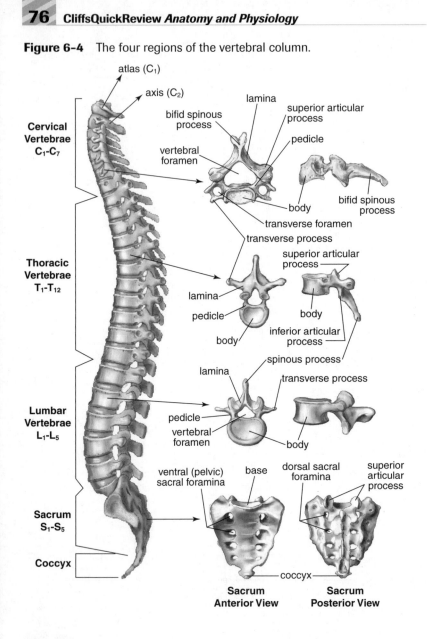

Thorax

The thoracic cage includes the thoracic vertebrae, sternum, ribs, and costal cartilages (see Figure 6-5). The sternum (breastbone) consists of three fused bones: the manubrium, body, and xiphoid process. There are 12 pairs of ribs, all of which attach to their posterior ends to vertebrae. At their anterior ends, they differ as to how they attach, as follows:

■ Seven pairs of true ribs (vertebrosternal ribs) attach directly to the sternum with hyaline cartilage called costal cartilage.

■ Three pairs of false ribs (vertebrochondral ribs) do not attach to the sternum. Rather, they connect (with costal cartilage) to the rib directly above them.

■ Two pairs of false ribs (floating ribs or vertebral ribs) do not attach to anything at their anterior ends.

Figure 6-5   The thoracic cage.

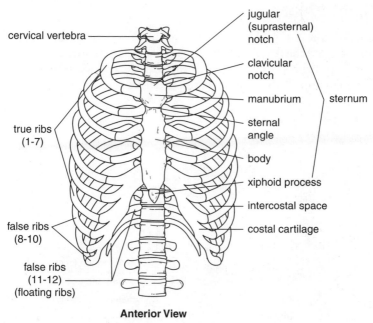

cervical vertebra

jugular (suprasternal) notch

clavicular notch

manubriun   sternum

true ribs (1-7)

sternal angle

body

xiphoid process

intercostal space

false ribs (8-10)

costal cartilage

false ribs (11-12) (floating ribs)

Anterior View

Here are important features of a rib:

■ The head is the end of the rib that articulates with the vertebral column.

■ The superior and inferior facets are facets on the head that articulate with the facets of the thoracic vertebrae.

- The neck, just beyond the head, bears a tubercle (rounded process) that articulates with the facet of the vertebral transverse process. Part of the tubercle also presents a place of attachment for ligaments.

- The costal angle designates the sharp turn of the rib.

- The costal groove, a passageway on the inside of the bending rib, provides for blood vessels and intercostals nerves.

- The body (shaft) is the major part of the rib—that part beyond the costal angle.

- Intercostal spaces, the areas between the ribs, are occupied by the intercostal muscles.

Pectoral Girdle

Each of the two pectoral (shoulder) girdles consists of two bones: the S-shaped clavicle and the flat, triangular scapula. The clavicle articulates with the sternum and the scapula. In turn, the scapula articulates with the humerus of the arm. Figure 6-6 illustrates details of these bones.

Figure 6-6 The pectoral girdle.

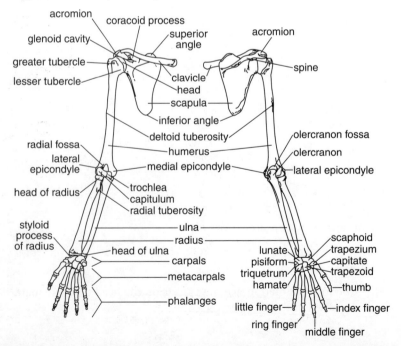

Upper Limb

The upper limb consists of the arm, forearm, and hand. The 30 bones of each upper limb are illustrated in Figure 6-6.

Pelvic Girdle

The pelvic (hip) girdle transfers the weight of the upper body to the legs. It consists of a pair of coxal bones (os coxae, hip bones), each of which contains three fused bones: the ilium, ishium, and pubis. Together with the sacrum and coccyx, the pelvic girdle forms a bowl-shaped region, the pelvis, that protects internal reproductive organs, the urinary bladder, and the lower part of the digestive tract. Features of these bones are given in Figure 6-7.

Figure 6-7 The pelvic girdle.

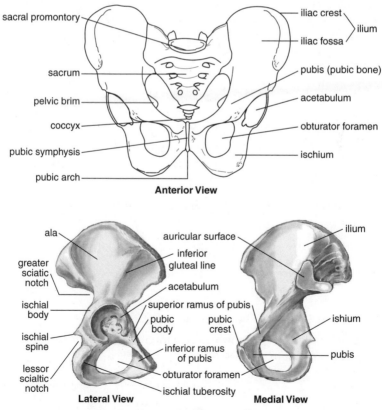

Lower Limb

The thigh, leg, and foot constitute the lower limb. The bones of the lower limbs are considerably larger and stronger than comparable bones of the upper limbs because the lower limbs must support the entire weight of the body while walking, running, or jumping. Figure 6-8 illustrates features of the 30 bones of each lower limb.

Figure 6-8 The 30 bones of each lower limb.

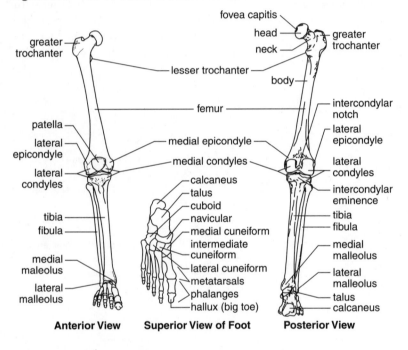

| Anterior View | Superior View of Foot | Posterior View |

Chapter Checkout

Q&A

1. A bone crucial to swallowing and speech is the _____ _____.

2. True or False. The sphenoid and occipital bones form most of the base of the cranium.

3. Which of the following bone (or bones) help(s) compose the face?
   a. the parietal bones
   b. the talus
   c. the zygomatic bone
   d. the vomer

4. True or False. The coccyx never varies in the number of vertebrae that compose it.

5. The outer, fibrocartilaginous ring of the intervertebral disk is called the _____ _____.

Answers: 1. hyoid bone 2. T 3. c & d 4. F 5. annulus fibrosus

Chapter 7

JOINTS

The intricate movements of a human, such as those performed in dance and athletics, are accomplished by using a wide variety of joints. Though joints allow the skeleton to be dynamic, they also play an important role in stability and protection. In fact, the mobility of a joint is often inversely proportional to its stability. For example, the sutures of the bones of the cranium are basically immovable in relationship to one another, but due to their stable nature, they serve to protect the brain throughout daily life and during incidents of trauma. On the other hand, the ball-and-socket of the shoulder enables a wide variety of complex movements. This increase in the amount of mobility leads to instability, which is why the shoulder is more susceptible to injury. The goal of this chapter is to familiarize you with the wide variety of joints that occur in the human body and to help you understand how these joints are classified.

Classifying Joints

A joint (articulation) occurs wherever bones meet. Joints are classified both structurally and functionally, as shown in Table 7-1.

Table 7-1   Joint Classification

Functional Class	Structural Class	Joint	Description Type	Example of Joint
Synarthrosis (immovable)	fibrous	suture	interlocking seams	between cranial bones
Synarthrosis (immovable)	fibrous	gom-phosis	peg-and-socket joint	between teeth & sockets
Synarthrosis (immovable)	cartilaginous	syn-chron-drosis	hyaline cartilage joint	between diaphysis & epiphysis in long bones
Amphiarthrosis (slightly movable)	fibrous	syndes-mosis	ligament ordistal interosseous membrane	joint of tibia & fibula
Amphiarthrosis (slightly movable)	cartilaginous	sym-physis	fibrocartilage acts as compressible cushion	interver-tebral discs of vertebral column
diarthrosis (freely movable)	synovial	gliding	two sliding surfaces	between carpals
diarthrosis (freely movable)	synovial	hinge	concave surface with convex surface	between humerus & ulna
diarthrosis (freely movable)	synovial	pivot	rounded end fits into ring of bone and ligament	between altas (C_1) & axis (C_2) vertebrae
diarthrosis (freely movable)	synovial	condyloid	oval condyle with oval cavity	between metacar-pals & pha-langes
diarthrosis (freely movable)	synovial	saddle	each surface is both concave and convex	between carpus & the first metacarpal
diarthrosis (freely movable)	synovial	ball- and-socket	ball-shaped head with cup-shaped socket	between femur & pelvis

Structural classification

Structural classification is based on the materials that hold the joint together and whether or not a cavity is present in the joint. There are three structural classes.

- Fibrous joints are held together by fibrous connective tissue. No joint cavity is present. Fibrous joints may be immovable or slightly movable.

- Cartilaginous joints are held together by cartilage (hyaline or fibrocartilage). No joint cavity is present. Cartilaginous joints may be immovable or slightly movable.

- Synovial joints are characterized by a synovial cavity (joint cavity) containing synovial fluid. Synovial joints are freely movable and characterize most joints of the body. Figure 7-1 lists other features of a synovial joint, including

  > Articular cartilage (hyaline cartilage), which covers the end of each bone.

  > A synovial membrane, which surrounds the synovial cavity. Its areolar connective tissue secretes a lubricating synovial fluid into the synovial cavity.

  > A fibrous capsule outside the synovial membrane, which surrounds the joint. It often contains bundles of dense, irregular, connective tissue called ligaments. The ligaments provide strength and flexibility to the joint.

  > The articulate capsule is composed of the synovial membrane and fibrous capsule.

  > Accessory ligaments lie outside the articular capsule (extracapsular ligaments) or inside the synovial cavity (intracapsular ligaments).

Functional classification

Functional classification is based on the degree to which the joint permits movement. There are three types:

- A synarthrosis joint permits no movement. Structurally, it may be a fibrous or cartilaginous joint.

- An amphiarthrosis joint permits only slight movement. Structurally, it may be fibrous or cartilaginous joint.

■ A diarthrosis joint is a freely movable joint. Structurally, it is always a synovial joint.

Figure 7-1 A synovial joint.

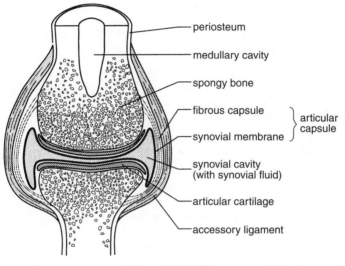

periosteum

medullary cavity

spongy bone

fibrous capsule

synovial membrane

} articular capsule

synovial cavity (with synovial fluid)

articular cartilage

accessory ligament

A Synovial Joint

Chapter Checkout

Q&A

1. Name the two types of joints where no joint cavity is present.
2. True or False. A diarthrosis joint allows little or no movement.
3. Which of the following is an example of an amphiarthrotic joint?
   a. between the femur and pelvis
   b. between the teeth and their sockets
   c. between cranial bones
   d. between the intervertebral disks
   e. between the metacarpals and phalanges

Critical Thinking Question

1. Discuss why each component of a synovial joint is critical to the function of the joint.

Answers: 1. fibrous and cartilaginous joints 2. F 3. d

Chapter 8

MUSCLE TISSUE

Chapter Check-In

❑ Discussing the structure and function of different muscle types

❑ Understanding the connective tissue associated with skeletal muscle

❑ Describing the physiology and stimulation of a muscular contraction

❑ Appreciating major metabolic pathways involved in cellular respiration

Muscle enables complex movements that are either under conscious control, such as turning the pages of this book, or involuntary, such as the contraction of the heart or the peristalsis in the gut. To understand how muscle accomplishes these various activities, you need to know the physiology behind a muscle contraction. This requires a detailed knowledge of the muscle's microscopic anatomy. Of course, muscle contractions will not take place without adequate nervous stimulation or a sufficient supply of ATP, the muscles' fuel. ATP is obtained via cellular respiration, which is accomplished by several different metabolic pathways discussed within this chapter.

Types of Muscles

There are three types of muscles:

■ Skeletal muscle is attached to bones and causes movements of the body. Skeletal muscle is also called striated muscle (because of its banding pattern when viewed under a microscope) or voluntary muscle (because muscle contraction can be consciously controlled).

■ Cardiac muscle is responsible for the rhythmic contractions of the heart. Cardiac muscle is involuntary—it generates its own stimuli to initiate muscle contraction.

■ Smooth muscle lines the walls of hollow organs. For example, it lines the walls of blood vessels and of the digestive tract where it serves to advance the movement of substances. Smooth muscle contraction is relatively slow and involuntary.

Connective Tissue Associated with Muscle Tissue

A skeletal muscle consists of numerous muscle cells called muscle fibers. Three layers of connective tissues surround these fibers to form a muscle. These and other connective tissues associated with muscles follow:

■ The endomysium is the connective tissue that surrounds each muscle fiber (cell).

■ The perimysium encircles a group of muscle fibers, forming a fascicle.

■ The epimysium encircles all the fascicles to form a complete muscle.

■ A tendon is a cordlike extension of the preceding three linings. It extends beyond the muscle tissue to connect the muscle to a bone or to other muscles.

■ An aponeurosis is a flat broad extension of the three muscle linings and serves the same function as a tendon.

■ The deep fascia surrounds the perimysium and encloses or lines other nearby structures such as blood vessels, nerves, and the body wall.

■ The superficial fascia (hypodermis or subcutaneous layer) lies immediately below the skin. The superficial fascia merges with the deep fascia where the surfaces of the skin meet.

Structure of Skeletal Muscle

A muscle fiber (cell) has special terminology and distinguishing characteristics:

■ The sarcolemma, or plasma membrane of the muscle cell, is highly invaginated by transverse tubules (T tubes) that permeate the cell.

■ The sarcoplasm, or cytoplasm of the muscle cell, contains calcium-storing sarcoplasmic reticulum, the specialized endoplasmic reticulum of a muscle cell.

■ Striated muscle cells are multinucleate. The nuclei lie along the periphery of the cell, forming swellings visible through the sarcolemma.

■ Nearly the entire volume of the cell is filled with numerous long myofibrils. Myofibrils consist of two types of filaments, shown in Figure 8-1:

> Thin filaments consist of two strands of the globular protein actin arranged in a double helix. Along the length of the helix are troponin and tropomyosin molecules that cover special binding sites on the actin.

> Thick filaments consist of groups of the filamentous protein myosin. Each myosin filament forms a protruding head at one end. An array of myosin filaments possesses protruding heads at numerous positions at both ends.

Figure 8-1 Two types of filaments.

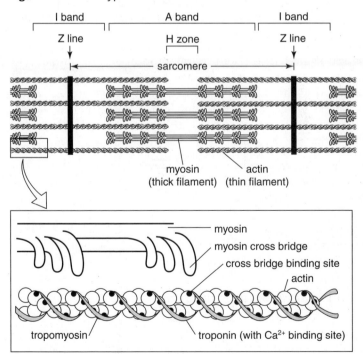

Within a myofibril, actin and myosin filaments are parallel and arranged side by side. The overlapping filaments produce a repeating pattern that gives skeletal muscle its striated appearance. Each repeating unit of the pattern, called a sarcomere, is separated by a border, or Z disc (Z line), to which the actin filaments are attached. The myosin filaments, with their protruding heads, float between the actin, unattached to the Z disc.

Muscle Contraction

Muscle contraction is described by the sliding-filaments model:

- ATP binds to a myosin head and forms ATP + P_i. When ATP binds to a myosin head, it is converted to ADP and P_i, which remain attached to the myosin head.

- Ca^{2+} exposes the binding sites on the actin filaments. Ca^{2+} binds to the troponin molecule causing tropomyosin to expose positions on the actin filament for the attachment of myosin heads.

- Cross bridges between myosin heads and actin filaments form. When attachment sites on the actin are exposed, the myosin heads bind to actin to form cross bridges.

- ADP and P_i are released, and sliding motion of actin results. The attachment of cross bridges between myosin and actin causes the release of ADP and P_i. This, in turn, causes a change in shape of the myosin head, which generates a sliding movement of the actin toward the center of the sarcomere. This pulls the two Z discs together, effectively contracting the muscle fiber to produce a power stroke.

- ATP causes the cross bridges to unbind. When a new ATP molecule attaches to the myosin head, the cross bridge between the actin and myosin breaks, returning the myosin head to its unattached position.

Without the addition of a new ATP molecule, the cross bridges remain attached to the actin filaments. This is why corpses become stiff with rigor mortis (new ATP molecules are unavailable).

Stimulation of muscle contraction

Neurons, or nerve cells, are stimulated when the polarity across their plasma membrane changes. The polarity change, called an action potential, travels along the neuron until it reaches the end of the neuron. A gap called a **synapse** or **synaptic cleft** separates the neuron from a muscle cell or another neuron. If a neuron stimulates a muscle, then the neuron is a motor neuron, and its specialized synapse is called a neuromuscular junction. Muscle contraction is stimulated through the following steps:

■ Action potential generates release of acetylcholine. When an action potential of a neuron reaches the neuromuscular junction, the neuron secretes the neurotransmitter acetylcholine (Ach), which diffuses across the synaptic cleft.

■ Action potential is generated on the motor end plate and throughout the T tubules. Receptors on the motor end plate, a highly folded region of the sarcolemma, initiate an action potential. The action potential travels along the sarcolemma throughout the transverse system of tubules.

■ Sarcoplasmic reticulum releases Ca^{2+}. As a result of the action potential throughout the transverse system of tubules, the sarcoplasmic reticulum releases Ca^{2+}.

■ Myosin cross bridges form. The Ca^{2+} released by the sarcoplasmic reticulum binds to troponin molecules on the actin helix, prompting tropomyosin molecules to expose binding sites for myosin cross-bridge formation. If ATP is available, muscle contraction begins.

Phases of a muscle contraction

A muscle contraction in response to a single nerve action potential is called a twitch contraction. A myogram, a graph of muscle strength (tension) with time, shows several phases, shown in Figure 8-2:

■ The latent period is the time required for the release of Ca^{2+}.

■ The contraction period represents the time during actual muscle contraction.

■ The relaxation period is the time during which Ca^{2+} is returned to the sarcoplasmic reticulum by active transport.

■ The refractory period is the time immediately following a stimulus during which the muscle fiber will not respond to a second stimulus. For skeletal muscle fibers, this period typically ends during the early part of the contraction period.

Figure 8-2 The phases of a myogram.

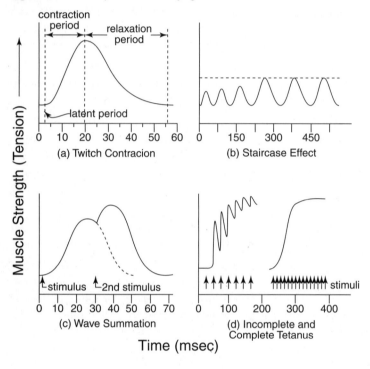

(a) Twitch Contracion

(b) Staircase Effect

(c) Wave Summation

(d) Incomplete and Complete Tetanus

Time (msec)

Quality of a muscle contraction

The following factors contribute to the strength and maximum duration of a muscle contraction:

■ Frequency of stimuli. If stimuli are repeatedly applied to a muscle fiber, Ca^{2+} may not be completely transported back into the sarcoplasmic reticulum before the next stimulus occurs. Depending upon the frequency of stimuli, Ca^{2+} may accumulate. In turn, the extra Ca^{2+} results in more power strokes and a stronger muscle contraction. Depending upon the frequency of stimuli, several effects are observed.

A staircase effect (treppe) is produced if each successive stimulus occurs after the relaxation period of the previous stimulus. Each successive muscle contraction is greater than the previous one, up to some maximum value. In addition to the accumulation of Ca^{2+}, other factors, such as increases in temperature and changes in pH, may contribute to this "warming up" effect commonly employed by athletes.

Wave (temporal) summation occurs if consecutive stimuli are applied during the relaxation period of each preceding muscle contraction. In this case, each subsequent contraction builds upon the previous contraction before its relaxation period ends.

Incomplete (unfused) tetanus occurs when the frequency of stimuli increases. Successive muscle contractions begin to blend, almost appearing as a single large contraction.

Complete (fused) tetanus occurs when the frequency of stimuli increases still further. In this case, individual muscle contractions completely fuse to produce one large muscle contraction.

- ■ Strength of stimulus. Muscle contractions intensify when more motor neurons stimulate more muscle fibers. This effect, called recruitment or multiple motor unit summation, is also responsible for fine motor coordination because by continually varying the stimulation of specific muscle fibers, smooth body movements are maintained.

- ■ Length of muscle fiber contraction. Because a muscle is attached to bones, muscle contraction is restricted to lengths that are between 60 percent and 175 percent of the length that produces optimal strength. This range of muscle lengths limits myosin cross bridges and actin only to positions where they overlap and thus can generate contractions.

- ■ Type of contraction. Muscle contraction implies that movement occurs between myosin cross bridges and actin. However, this movement does not necessarily result in shortening of the muscle. As a result, two kinds of muscle contractions are defined.

    Isotonic contractions occur when muscles change length during a contraction. Picking up a book is an example.

    Isometric contractions occur when muscles do not change length during a contraction. When holding a book in midair, muscle fibers produce a force, but no motion is generated.

- ■ Type of muscle fiber. Muscle fibers are classified into two groups.

    Slow fibers contract slowly, have a high endurance, and are red from their rich blood supply. However, they do not produce much strength. These fibers are used for long distance running.

    Fast fibers contract rapidly, fatigue rapidly, and are white because the blood supply is limited. They generate considerable strength. These fibers are used for short distance running.

- Muscle tone. In any relaxed skeletal muscle, a small number of contractions continuously occur. Observed as firmness in a muscle, these contractions maintain body posture and increase muscle readiness.

- Muscle fatigue. Muscle fibers stop contraction when inadequate amounts of ATP are available. Lack of oxygen and glycogen and the accumulation of lactic acid (a byproduct of ATP production in the absence of oxygen), together with the lack of ATP, all contribute to muscle fatigue.

Muscle metabolism

In order for muscles to contract, ATP must be available in the muscle fiber. ATP is available from the following sources:

- ATP from within the cell. ATP available within the muscle fiber can maintain muscle contraction for several seconds.

- ATP from creatine phosphate. Creatine phosphate, a high-energy molecule stored in muscle cells, transfers its high-energy phosphate group to ADP to form ATP. The creatine phosphate in muscle cells is able to generate enough ATP to maintain muscle contraction for about 15 seconds.

- ATP from glucose stored within the cell. Glucose within the cell is stored in the carbohydrate glycogen. Through the metabolic process of glycogenolysis, glycogen is broken down to release glucose. ATP is then generated from glucose by cellular respiration.

- ATP from glucose and fatty acids obtained from the bloodstream. When energy requirements are high, glucose from glycogen stored in the liver and fatty acids from fat stored in adipose cells and the liver are released into the bloodstream. Glucose and fatty acids are then absorbed from the bloodstream by muscle cells. ATP is then generated from these energy-rich molecules by cellular respiration.

Cellular respiration is the process by which ATP is obtained from energy-rich molecules. Several major metabolic pathways are involved, some of which require the presence of oxygen. Here's a summary of the important pathways:

- In glycolysis, glucose is broken down to pyruvic acid, and two ATP molecules are generated. Because no oxygen is used during the various metabolic steps of this pathway, glycolysis is called an anaerobic process.

- In anaerobic respiration, pyruvic acid (from glycolysis) is converted to lactic acid. No ATP is generated and, as its name indicates, no oxygen is required. The importance of this process is that it regenerates certain coenzymes necessary for glycolysis to continue. Thus, in the absence of oxygen, anaerobic respiration is indirectly responsible for the production of two ATPs (during glycolysis).

   Advantages of anaerobic respiration: Anaerobic respiration is relatively rapid, and it does not require oxygen.

   Disadvantages of anaerobic respiration: Anaerobic respiration generates only two ATPs, and lactic acid is produced. Most lactic acid diffuses out of the cell and into the bloodstream and is subsequently absorbed by the liver. Some of the lactic acid remains in the muscle fibers, where it contributes to muscle fatigue. Because both the liver and muscle fibers must convert the lactic acid back to pyruvic acid when oxygen becomes available, anaerobic respiration is said to produce oxygen debt.

- In aerobic respiration, pyruvic acid (from glycolysis) and fatty acids (from the bloodstream) are broken down, producing H_2O and CO_2 (carbon dioxide) and regenerating the coenzymes for glycolysis. A total of 36 ATP molecules is produced (including the two from glycolysis). However, oxygen is required for this pathway.

   Advantages of aerobic respiration: Aerobic respiration generates a large amount of ATP.

   Disadvantages of aerobic respiration: Aerobic respiration is relatively slow and requires oxygen.

When the ATP generated from creatine phosphate is depleted, the immediate requirements of contracting muscle fibers force anaerobic respiration to begin. Anaerobic respiration can supply ATP for about 30 seconds. If muscle contraction continues, aerobic respiration, the slower ATP-producing pathway, begins and produces large amounts of ATP as long as oxygen is available. Eventually, oxygen is depleted, and aerobic respiration stops. However, ATP production by anaerobic respiration may still support some further muscle contraction. Ultimately, the accumulation of lactic acid from anaerobic respiration and the depletion of resources (ATP, oxygen, and glycogen) lead to muscle fatigue, and muscle contraction stops.

Structure of Cardiac Muscle

Although it is striated, cardiac muscle differs from skeletal muscle in that it is highly branched with cells connected by overlapping projections of the sarcolemma called intercalated discs. These discs contain desmosomes and gap junctions. In addition, cardiac muscle is autorhythmic, generating its own action potential, which spreads rapidly throughout muscle tissue by electrical synapses across the gap junctions.

Structure of Smooth Muscle

Due to its irregular arrangement of actin and myosin filaments, smooth muscle does not have the striated appearance of skeletal muscle. In addition, the sarcolemma does not form a system of transverse tubules. As a result, contraction is controlled and relatively slow; properties appropriate for smooth muscle function.

In addition to the thick myosin and thin actin filaments, smooth muscles also possess noncontracting intermediate filaments. The intermediate fibers attach to dense bodies that are scattered through the sarcoplasm and attached to the sarcolemma. During contraction, the movement of myosin and actin is transferred to intermediate fibers, which pull on the dense bodies, which, in turn, pull the muscle cell together. In this way, the dense bodies function similarly to the Z discs in striated muscles.

Chapter Checkout

Q&A

1. In anaerobic respiration, _____ _____ accumulates and leads to muscle fatigue.

2. True or False. Fast-twitch fibers contract rapidly, fatigue rapidly, and are highly vascularized.

3. Which of the following is true of glycolysis?
   a. pyruvic acid is broken down to produce ATP
   b. cellular respiration primarily utilized in a 100-meter dash
   c. is the slowest form of cellular respiration
   d. produces oxygen debt
   e. oxygen is vital for this type of cellular respiration

4. True or False. In muscle contractions, the length of the muscle is always shortened.

5. The _____ _____ is the time immediately following a stimulus where the muscle will not respond to a second stimulus.

Answers: 1. lactic acid 2. F 3. b 4. F (isometric contractions) 5. refractory period

Chapter 9

THE MUSCULAR SYSTEM

Chapter Check-In

❑ Characterizing how muscles are named

❑ Identifying the muscles associated with each region of the body

❑ Listing the various muscles of the body

Knowing the muscular organization of each region of the body is crucial in anatomy. With an understanding of where a muscle originates and inserts, you can calculate the movements that will occur at a joint when these two points are brought together following an isotonic muscular contraction. The orientation, placement, and coordination of these muscles allow the human body to produce a wide range of voluntary movements.

The **muscular system** consists of skeletal muscles and their associated connective tissues. It does not include cardiac muscle or smooth muscle, which are associated with the systems in which they are found, such as the cardiovascular, digestive, urinary, or other organ systems.

Skeletal Muscle Actions

A skeletal muscle may attach a bone to another bone (often across a joint) or a bone to another structure, such as skin. When the muscle contracts, one of the structures usually remains stationary, while the other moves. The following terms refer to this characteristic of muscle contraction.

■ The origin of the muscle is the muscle end that attaches to the stationary structure, usually bone.

■ The **insertion** of the muscle is the muscle end that attaches to the moving structure.

- The **belly** of the muscle is that part of the muscle between the origin and insertion.

Several muscles usually influence a particular body movement:

- The prime mover is the muscle that is most responsible for the movement.

- Synergists are other muscles that assist the prime mover. Synergists may stabilize nearby bones or refine the movement of the prime mover.

- Antagonists are muscles that cause a movement opposite to that of the prime mover. If the prime mover raises an arm, then its antagonist pulls the arm down. An antagonist is generally attached to the opposite side of the joint to which the prime mover is attached.

Names of Skeletal Muscles

Skeletal muscles are often named after the following characteristics:

- Number of origins. Biceps, triceps, and quadriceps indicate two, three, and four origins, respectively.

- Location of origin or insertion. The sternocleidomastoid names the sternum ("sterno") and clavicle ("cleido") as its origins and the mastoid process of the temporal bone as its insertion.

- Location. In addition to its origin or insertion, a muscle name may indicate a nearby bone or body region. For example, the temporalis muscle covers the temporal bone.

- Shape. The deltoid (triangular), trapezius (trapezoid), serratus (saw-toothed) and rhomboideus major (rhomboid) muscles have names that describe their shapes.

- Direction of muscle fibers. The terms rectus (parallel), transverse (perpendicular), and oblique (at an angle) in muscle names refer to the direction of the muscle fibers with respect to the midline of the body.

- Size. Maximus (largest), minimus (smallest), longus (longest), and brevis (shortest) are common suffixes added to muscle names.

■ Action. Terms such as flexor, extensor, abductor, and adductor are added as prefixes to muscle names to indicate the kind of movement generated by the muscle.

Muscle Size and Arrangement of Muscle Fascicles

The size of a muscle influences its capabilities. When a muscle fiber (cell) contracts, it can shorten to nearly half its relaxed length. The longer a muscle fiber, then the greater range of movement it can generate. In contrast, an increase in the number of muscle fibers increases the strength of the contraction.

Muscle fibers are grouped into fascicles, which are, in turn, grouped together to form a muscle. The size (length) and number of fascicles determine the strength and range of movement of a muscle. Common fascicle patterns follow:

■ Parallel fascicles have their long axes parallel to each other. Parallel fascicles can be flat, or straplike, or they can bulge at their bellies and be spindle shaped, or fusiform.

■ Circular fascicles are arranged in concentric rings. Muscles with this pattern form sphincter muscles that control the opening and closing of orifices.

■ Pennate fascicles are short and attach obliquely to a long tendon that extends across the entire muscle. In a unipennate pattern, the muscle resembles one half of a feather (the tendon is represented by the shaft of the feather). A bipennate pattern resembles a complete feather, with fascicles attached to both sides of a central tendon. A multipennate pattern of fascicles resembles three or more feathers attached at their bases.

Major Skeletal Muscles

The major skeletal muscles are described in Tables 9-1 through 9-5 and illustrated in Figures 9-1 through 9-6.

Figure 9-1 The major skeletal muscles—anterior superficial view.

Anterior Superficial View

Figure 9-2 The major skeletal muscles—posterior superficial view.

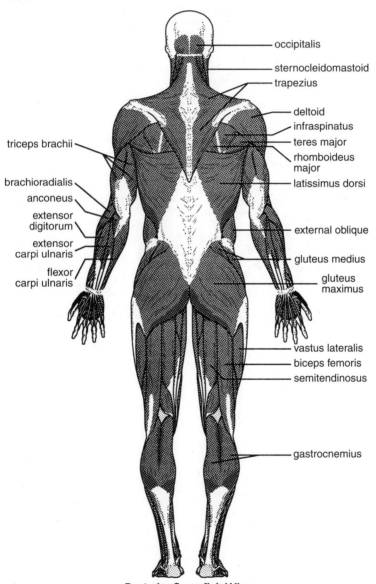

Posterior Superficial View

Table 9-1   Muscles of the Head and Neck

Muscle	*Origin/Insertion*	*Action*
Epicranius:frontalis	O: galea aponeurotica I: skin around eyes	raises eyebrows; surprised
Epicranius:occiptalis	O: occipital bone I : galea aponeurotica	pulls scalp back; surprised
Orbicularis oculi	O: maxillary & frontal bones I: eyelids	closes eyelids; blinking
Orbicularis oris	O: muscle fibers around mouth I: skin around mouth	closes lips; kissing
Buccinator	O: maxilla & mandible I: orbicularis oris	compresses cheek; whistling
Platysma	O: fascia in upper chest I: mandible & corner of mouth	lowers mandible; opens mouth
Mentalis	O: mandible I : skin of chin	protrudes lower lip; pouting
Risorius	O: fascia on masseter muscle I: skin at corner of mouth	lateral movement of lips; grimacing
Zygomaticus	O: zygomatic bone I: skin around mouth	raises edges of mouth; smiling
Levator labi superioris	O: infraorbital margin of maxilla I: skin of upper lip	raises upper lip; as in disgust
Depressor labi inferioris	O: mandible I: skin of lower lip	lowers lower lip
Temporalis	O: parietal bone I: mandible	raises mandible; closes mouth
Masseter	O: zygomatic arch I: mandible	raises mandible; closes mouth
Medial pterygoid	O: sphenoid & maxilla I: mandible	raises mandible; side-to-side mouth motion
Lateral pterygoid	O: sphenoid & maxilla I: mandible	raises mandible; side-to side mouth motion
Sternocleidomastoid	O: ternum & clavicle I: temporal bone	flexes & rotates head

(continued)

Table 9-1 *(continued)*

Muscle	Origin/Insertion	Action
Splenius capitis	O: cervical & thoracic vertebrae I: temporal bone	rotates, bends, or extends head
Semispinalis capitis	O: cervical & thoracic vertebrae I: occipital bone	rotates or extends head
Longissimus	O: cervical & thoracic vertebrae I: temporal bone	rotates, bends, or extends head
Omohyoid	O: scapula I: hyoid bone	depresses hyoid bone
Sternohyoid	O: sternum & clavicle I: hyoid bone	depresses hyoid bone

Table 9-2   Muscles of the Neck, Shoulder, Thorax, and Abdominal Wall

Muscle	Origin/Insertion	Action
Semispinalis capitis	O: cervical & thoracic vertebrae I: occipital bone	extends & rotates head
Splenius capitis	O: c & t vertebrae I : occipital & temporal bone	extends and rotates head
Deltoid	O: clavicle & scapula I: humerus	abducts, flexes, extends, & rotates arm
Pectoralis major	O: clavicle, sternum, ribs I: humerus	flexes, adducts, & rotates arm
Infraspinatus	O: scapula I: humerus	rotates arm
Teres major	O: scapula I: humerus	extends, rotates arm
Latissimus dorsi	O: vertebrae, ribs, ilium I : humerus	extends, adducts, rotates arm

Muscle	Origin/Insertion	Action
Levator scapulae	O: cervical vertebrae I: scapula	elevates scapula
Pectoralis minor	O: ribs I: scapula	stabilizes scapula, elevates ribs
Serratus anterior	O: ribs I: scapula	stabilizes scapula, elevates ribs
Trapezius	O: occipital bone & c & t vertebrae; I : scapula & clavicle	elevates, adducts, & rotates scapula
Rhomboideous major/Rhomboideus minor	O: c & t vertebrae I: scapula	adducts & rotates scapula
Rectus abdominis	O: pubic crest & symphysis I: xiphoid process & ribs	flexes vertebral column, compresses abdomen
External oblique	O: ribs I : linea alba, ilium	compresses abdomen, rotates trunk
Transverse abdominis	O: ilium, ribs I: linea alba, xiphoid process	compresses abdomen
External intercostals	O: lower border of rib above I: upper border of rib below	elevates ribs, aids inspiration
Internal intercostals	O: upper border of rib below I: lower border of rib above	pulls ribs together, aids expiration
Diaphragm	O: lower ribs, sternum I: central tendon	aids inspiration
Spinalis	O: lumbar & thoracic vertebrae I: thoracic & cervical vertebrae	extends vertebral column
Longissimus	O: lumbar & cervical vertebrae I : temporal bone, vertebrae	extends vertebral column
Iliocostalis	O: ilium, ribs I: ribs	extends vertebral column

Figure 9-3   The major skeletal muscles—lateral view.

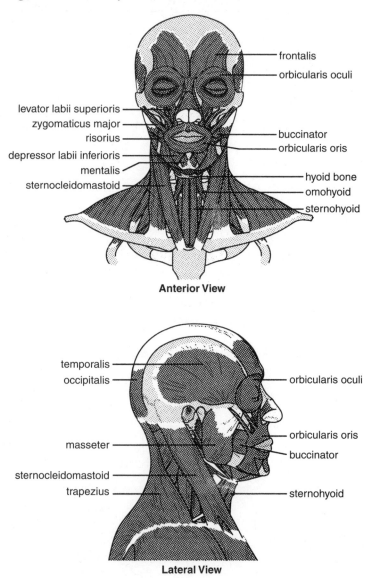

Anterior View

Lateral View

Figure 9-4 The major skeletal muscles—anterior superficial view, anterior deep view, posterior superficial view, and posterior deep view.

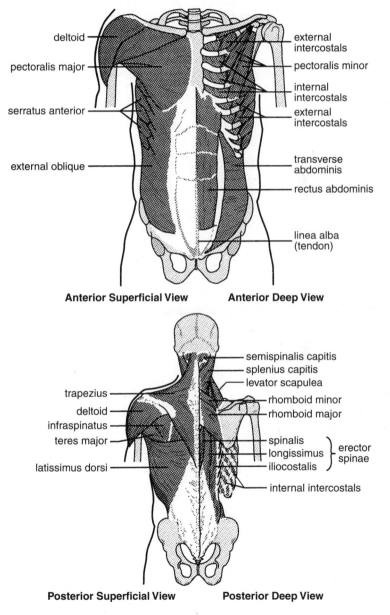

Anterior Superficial View   **Anterior Deep View**

Posterior Superficial View   **Posterior Deep View**

Table 9-3   Muscles of the Arm and Forearm

Muscle	Origin/Insertion	Action
Coracobrachiolis	O: scapula I: humerus	flexes & adducts arm
Biceps brachii	O: scapula, glenoid cavity I: radius	flexes arm, flexes forearm, & rotates hand
Brachialis	O: humerus I: ulna	flexes forearm
Brachioradialis	O: humerus I: radius	flexes forearm
Triceps brachii	O: humerus I: ulna	extends forearm
Anconeus	O: humerus I: ulna	extends forearm
Pronator teres	O: humerus, ulna I: radius	rotatesforearm
Pronator quadratus	O: ulna I: radius	rotates forearm
Supinator	O: ulna I: radius	rotates forearm
Flexor carpi radialis	O: humerus I : metacarpals	flexes & abducts wrist
Flexor carpi ulnaris	O : humerus, ulna I : carpals, metacarpals	flexes & abducts wrist
Flexor digitorum superficialis	O: humerus, ulna, radius I: phalanges	flexes finger 2 - 5
Flexor digitorum profundus	O: ulna I: phalanges	flexes distal fingers 2 - 5
Palmaris longus	O: humerus I : flexor retinaculum	flexes wrist
Extensor carpi radialis longus	O : humerus I : second metacarpal	extends & abducts wrist
Extensor carpi ulnaris	O: humerus, ulna I: fifth metacarpal	extends & adducts wrist
Extensor digitorum	O: humerus I: distal phalanges	extends fingers 2 - 5
Extensor pollicis brevis	O: radius I: phalanx of thumb	extends thumb
Extensor pollicis longus	O: radius I: phalanx of thumb	extends thumb
Extensor indicis	O: ulna I: index finger	extends index finger
Abductor pollicis longus	O: radius & thumb I: first metacarpal	abducts & extends thumb

Table 9-4 Muscles of the Thigh and Leg

Muscle	Origin/Insertion	Action
Gluteus maximus	O: ilium, sacrum ,coccyx I: femur	extends & rotates thigh
Gluteus medius	O: ilium I: femur	abducts & rotates thigh
Pectineus	O: pubis I: femur	adducts & flexes thigh
Adductor longus	O: pubis I: femur	adducts, flexes, & rotates thigh
Adductor brevis	O: pubis I: femur	adducts, flexes, & rotates thigh
Adductor magnus	O: pubis, ischium I: femur	adducts, flexes, & rotates thigh
Gracilis	O: pubis I: tibia	adducts thigh & flexes leg
Sartorius	O: ilium I: tibia	flexes & rotates thigh, flexes leg
Quadriceps femoris: rectus femoris	O: ilium, femur I: patella, tibia	extends leg, flexes thigh
Quadriceps femoris: vastus lateralis	O: ilium, femur I: patella, tibia	extends leg
Quadriceps femoris: vastus medialis	O: ilium, femur I: patella, tibia	extends leg
Quadriceps femoris: vastus intermedius	O: ilium, femur I: patella, tibia	extends leg
Biceps femoris (hamstrings)	O: ischium, femur I: fibula	flexes & rotates leg, extends thigh
Semitendinosus (hamstrings)	O: ischium I: tibia	flexes & rotates leg, extends thigh
Semimembranosus	O: ischium I: tibia	flexes & rotates leg, extends thigh
Tibialis anterior	O: tibia I: 1st metatarsal & cuneiform	dorsiflexes & inverts foot
Extensor digitorum longus	O: tibia, fibula I: phalanges of toes	dorsiflexes & everts foot, extends toes
Gastrocnemius	O: femur I: calcanues	plantar flexes foot, flexes leg
Soleus	O: tibia, fibula I: calcaneus	plantar flexes foot

Figure 9-5 The major skeletal muscles—anterior superficial view, anterior deep view, posterior superficial view, and posterior deep view.

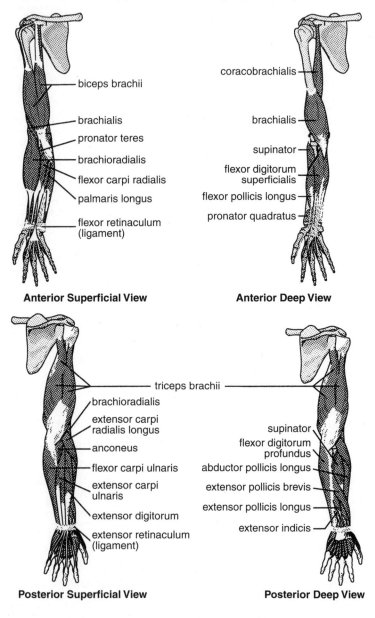

biceps brachii

brachialis

pronator teres

brachioradialis

flexor carpi radialis

palmaris longus

flexor retinaculum
(ligament)

Anterior Superficial View

coracobrachialis

brachialis

supinator

flexor digitorum
superficialis

flexor pollicis longus

pronator quadratus

Anterior Deep View

triceps brachii

brachioradialis

extensor carpi
radialis longus

anconeus

flexor carpi ulnaris

extensor carpi
ulnaris

extensor digitorum

extensor retinaculum
(ligament)

Posterior Superficial View

supinator

flexor digitorum
profundus

abductor pollicis longus

extensor pollicis brevis

extensor pollicis longus

extensor indicis

Posterior Deep View

Figure 9-6 The major skeletal muscles—anterior superficial view, anterior deep view, posterior superficial view, and posterior deep view.

pectineus
adductor longus
sartorius
gracilis
rectus femoris
vastus lateralis
vastus medialis
} quadriceps femoris
tibialis anterior
extensor digitorum longus

adductor brevis
adductor magnus

Anterior Superficial View   **Anterior Deep View**

gluteus medius
gluteus maximus
adductor magnus
gracilis
vastus lateralis
biceps femoris
semitendinosus
gastrocnemius

soleus

Posterior Superficial View   **Posterior Deep View**

Chapter Checkout

Q&A

1. Muscles that cause a movement opposite to the prime mover are called _____.

2. True or False. The omohyoid muscle originates from the scapula.

3. Which of the following muscle(s) is/are important in facial expression?

   a. risorius

   b. teres major

   c. sternocleidomastoid

   d. occipitalis

4. Which of the following muscle(s) adduct(s) the thigh?

   a. sartorius
   b. pectineus
   c. gracilis
   d. adductor pollicis longus

Critical Thinking Questions

1. Can you name the muscles that either originate or insert on the scapula?

Answers: 1. antagonists 2. T 3. a & d 4. b & c

Chapter 10

NERVOUS TISSUE

Chapter Check-In

❑ Naming the basic parts of a neuron

❑ Classifying neurons according to function and structure

❑ Listing the different types of neuroglia that support and protect neurons

❑ Describing how a neural impulse is generated and moves across a synapse

The human nervous system accomplishes countless actions that are crucial to existence. The functional cellular unit of this complex system is the neuron. Moreover, the primary goal of neuroscience is to understand the complexities of the nervous system in terms of the morphology and physiology of the neuron and its support cells, the neuroglia.

Nervous tissue consists of two kinds of nerve cells—neurons and neuroglia.

Neurons

A **neuron** is a cell that transmits nerve impulses. It consists of the following parts, shown in Figure 10-1.

■ The cell body (soma or perikaryon) contains the nucleus and other cell organelles.

■ Nissl bodies (chromatophilic substance) are clusters of rough endoplasmic reticulum, sites of protein synthesis.

■ The dendrite is typically a short, abundantly branched, slender process (extension) of the cell body that receives stimuli.

■ The axon is typically a long, slender process of the cell body that sends nerve impulses. It emerges from the cell body at the cone-shaped axon hillock. Nerve impulses arise in the trigger zone, generally located in the initial segment, an area just outside the axon hillock. The cytoplasm of the axon, the axoplasm, is surrounded by its plasma membrane, the axolemma. A few axons branch along their lengths to form axon collaterals, and these branches may return to merge with the main axon. At its end, each axon or axon collateral usually forms numerous branches (telodendria), with most branches terminating in bulb-shaped structures called synaptic knobs (synaptic end bulbs). The synaptic knobs contain neurotransmitters, chemicals that transmit nerve impulses to a muscle or another neuron.

Figure 10-1 Parts of a neuron.

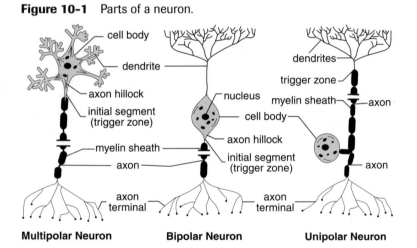

Neurons can be classified by function or by structure. Functionally, they fall into three groups:

■ Sensory neurons (afferent neurons) transmit sensory impulses from the skin and other sensory organs or from various places within the body toward the central nervous system (CNS, or brain and spinal cord).

■ Motor neurons (efferent neurons) transmit nerve impulses from the CNS toward effectors, target cells that produce some kind of response. Effectors include muscles, sweat glands, and exocrine glands.

■ Association neurons (interneurons) are located in the CNS and transmit impulses from sensory neurons to motor neurons. More than 90 percent of the neurons of the body are association neurons.

Neurons are structurally classified into three groups, as shown in Figure 10-1:

■ Multipolar neurons have one axon and several to numerous dendrites. Most neurons are of this type.

■ Bipolar neurons have one axon and one dendrite. They emerge from opposite sides of the cell body. Bipolar neurons are found only as specialized sensory neurons in the eye, ear, or olfactory organs.

■ Unipolar neurons have one process of emerging from the cell body that branches, T-fashion, into two processes. Both processes function together as a single axon. Dendrites emerge from one of the terminal ends of the axon. The trigger zone in a unipolar neuron is located at the junction of the axon and dendrites. Unipolar neurons are mostly sensory neurons.

The following terms apply to neurons and groups of neurons:

■ A nerve fiber is an axon or dendrite.

■ A **nerve** is a bundle of nerve fibers in the peripheral nervous system (PNS, or outside the CNS) held together with connective tissue. Most nerves contain both sensory and motor fibers. Cell bodies are usually grouped into separate bundles called ganglia.

■ A nerve tract is a bundle of nerve fibers in the CNS.

Neuroglia

Neuroglia (glia) are cells that support and protect neurons. The following four neuroglia are found in the CNS:

■ Astrocytes have numerous processes that give the cell a star-shaped appearance. Astrocytes maintain the ion balance around neurons and control the exchange of materials between blood vessels and neurons.

■ Oligodendrocytes have fewer processes than astrocytes. They wrap these cytoplasmic processes around neurons to create an insulating barrier called a myelin sheath.

■ Microglia are phagocytic macrophages that provide a protective function by engulfing microorganisms and cellular debris.

■ Ependymal cells line the fluid-filled cavities of the brain and spinal cord. Many are ciliated.

Two kinds of neuroglia are found in PNS:

■ Schwann cells (neurolemmocytes) wrap around axons to produce an insulating myelin sheath. Schwann cells provide the same function in the PNS as oligodendrocytes provide in the CNS.

■ Satellite cells are located in ganglia where they surround the cell bodies of neurons.

Myelination

In the PNS, the myelin sheath is an insulation formed by Schwann cells around axons (refer to Figure 10-1). Each Schwann cell tightly wraps around the axon numerous times to form a multilayered insulation. The last wrapping of the plasma membrane of the Schwann cell, the neurilemma, is loose and contains the nucleus, cytoplasm, and organelles of the Schwann cell.

The myelin sheath consists of numerous Schwann cell wrappings along the length of the axon. Spaces occur between adjacent Schwann cells, leaving uninsulated areas, or neurofibral nodes (nodes of Ranvier), along the axons. As an insulator, the Schwann cells interrupt the continuous conduction of a nerve impulse along the axon. In response, the nerve impulse jumps across each Schwann cell, from one neurofibral node to the next (saltatory conduction). In this fashion, the myelin sheath provides insulation between adjacent nerve fibers, preventing the crossover of one nerve impulse to an adjacent axon.

In the CNS, the axons of neurons are insulated by oligodendrocytes. The myelin sheath is formed when a process extended by the oligodendrocyte wraps around an axon. Many axons may be myelinated by multiple processes from a single oligodendrocyte. The oligodendrocyte does not form a neurilemma because its cell body does not wrap around any axons.

The white and gray matter of the brain and spinal cord are distinguished by the presence or absence of myelin sheaths.

■ White matter contains the myelinated axons of neurons. (The white color is from the myelin sheaths.)

■ Gray matter contains the unmyelinated portions of neurons (cell bodies, dendrites, and axon terminals), unmyelinated neurons, and neuroglia.

Transmission of Nerve Impulses

The transmission of a nerve impulse along a neuron from one end to the other occurs as a result of chemical changes across the membrane of the neuron. The membrane of an unstimulated neuron is polarized—that is, there is a difference in electrical charge between the outside and inside of the membrane. The inside is negative with respect to the outside.

Polarization is established by maintaining an excess of sodium ions (Na^+) on the outside and an excess of potassium ions (K^+) on the inside. A certain amount of Na^+ and K^+ is always leaking across the membrane through leakage channels, but Na^+/K^+ pumps in the membrane actively restore the ions to the appropriate side.

Other ions, such as large, negatively charged proteins and nucleic acids, reside within the cell. It is these large, negatively charged ions that contribute to the overall negative charge on the inside of the cell membrane as compared to the outside.

In addition to crossing the membrane through leakage channels, ions may also cross through gated channels. Gated channels open in response to neurotransmitters, changes in membrane potential, or other stimuli.

The following events characterize the transmission of a nerve impulse (see Figure 10-2:

- **Resting potential.** The resting potential describes the unstimulated, polarized state of a neuron (at about -70 millivolts).

- **Graded potential.** A graded potential is a change in the resting potential of the plasma membrane in the response to a stimulus. A graded potential occurs when the stimulus causes Na^+ or K^+ gated channels to open. If Na^+ channels open, positive sodium ions enter, and the membrane depolarizes (becomes more positive). If the stimulus opens K^+ channels, then positive potassium ions exit across the membrane and the membrane hyperpolarizes (becomes more negative). A graded potential is a local event that does not travel far from its origin. Graded potentials occur in cell bodies and dendrites. Light, heat, mechanical pressure, and chemicals, such as neurotransmitters, are examples of stimuli that may generate a graded potential (depending upon the neuron).

- **Action potential.** Unlike a graded potential, an action potential is capable of traveling long distances. If a depolarizing graded potential is sufficiently large, Na^+ channels in the trigger zone open. In response,

Na⁺ on the outside of the membrane becomes depolarized (as in a graded potential). If the stimulus is strong enough—that is, if it is above a certain threshold level—additional Na⁺ gates open, increasing the flow of Na⁺ even more, causing an action potential, or complete depolarization (from -70 to about +30 millivolts). This, in turn, stimulates neighboring Na⁺ gates, farther down the axon, to open. In this manner, the action potential travels down the length of the axon as opened Na⁺ gates stimulate neighboring Na⁺ gates to open. The action potential is an all-or-nothing event: When the stimulus fails to produce depolarization that exceeds the threshold value, no action potential results, but when threshold potential is exceeded, complete depolarization occurs.

Figure 10-2 Events that characterize the transmission of a nerve impulse.

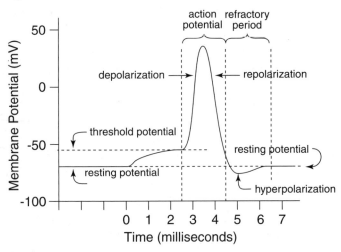

- Repolarization. In response to the inflow of Na⁺, K⁺ channels open, this time allowing K⁺ on the inside to rush out of the cell. The movement of K⁺ out of the cell causes repolarization by restoring the original membrane polarization. Unlike the resting potential, however, in repolarization the K⁺ are on the outside and the Na⁺ are on the inside. Soon after the K⁺ gates open, the Na⁺ gates close.

- Hyperpolarization. By the time the K⁺ channels close, more K⁺ have moved out of the cell than is actually necessary to establish the original polarized potential. Thus, the membrane becomes hyperpolarized (about -80 millivolts).

■ Refractory period. With the passage of the action potential, the cell membrane is in an unusual state of affairs. The membrane is polarized, but the Na^+ and K^+ are on the wrong sides of the membrane. During this refractory period, the axon will not respond to a new stimulus. To reestablish the original distribution of these ions, the Na^+ and K^+ are returned to their resting potential location by Na^+/K^+ pumps in the cell membrane. Once these ions are completely returned to their resting potential location, the neuron is ready for another stimulus.

The Synapse

A synapse, or synaptic cleft, is the gap that separates adjacent neurons or a neuron and a muscle. Transmission of an impulse across a synapse, from presynaptic cell to postsynaptic cell, may be electrical or chemical. In electrical synapses, the action potential travels along the membranes of gap junctions, small tubes of cytoplasm along the membranes of gap junctions, small tubes of cytoplasm that allow the transfer of ions between adjacent cells. In chemical synapses, action potentials are transferred across the synapse by the diffusion of chemicals, as follows:

■ Calcium (Ca^{2+}) gates open. When an action potential reaches the end of an axon, the depolarization of the membrane causes gated channels to open that allow Ca^{2+} to enter.

■ Synaptic vesicles release neurotransmitter. The influx of Ca^{2+} into the terminal end of the axon causes synaptic vesicles to merge with the presynaptic membrane, releasing a neurotransmitter into the synaptic cleft.

■ Neurotransmitter binds with postsynaptic receptors. The neurotransmitter diffuses across the synaptic cleft and binds with specialized protein receptors on the postsynaptic membrane. Different proteins are receptors for different neurotransmitters.

■ The postsynaptic membrane is excited or inhibited. Depending upon the kind of neurotransmitter and the kind of membrane receptor, there are two possible outcomes for the postsynaptic membrane, both of which are graded potentials.

If positive ion gates open (which allow more Na^+ and Ca^{2+} to enter than K^+ to exit), the membrane becomes depolarized, which results in an excitatory postsynaptic potential (EPSP). If the threshold potential is exceeded, an action potential is generated.

If K$^+$ or chlorine ion (Cl$^-$) gates open (allowing K$^+$ to exit or Cl$^-$ to enter), the membrane becomes more polarized (hyperpolarized), which results in an inhibitory postsynaptic potential (IPSP). As a result, it becomes more difficult to generate an action potential on this membrane.

■ The neurotransmitter is degraded and recycled. After the neurotransmitter binds to the postsynaptic membrane receptors, it is either transported back to and reabsorbed by the secreting neuron, or it is broken down by enzymes in the synaptic cleft. For example, the common neurotransmitter acetylcholine is broken down by cholinesterase. Reabsorbed and degraded neurotransmitters are recycled by the presynaptic cell.

Here are some of the common neurotransmitters and the kinds of activity they generate:

■ Acetylcholine (ACh) is commonly secreted at neuromuscular junctions, the gaps between motor neurons and muscle cells, where it stimulates muscles to contract (by opening gated positive ion channels). At other kinds of junctions, it typically produces an inhibitory postsynaptic potential.

■ Epinephrine, norepinephrine (NE), dopamine, and serotonin are derived from amino acids and are secreted mostly between neurons of the CNS.

■ Gamma aminobutyric acid (GABA) is usually an inhibitory neurotransmitter (opening gated Cl$^-$ channels) among neurons in the brain.

Chapter Checkout

Q&A

1. _____ neurons transmit nerve impulses away from the central nervous system (CNS).

2. True or False. Gray matter contains the myelinated axons of neurons.

3. True or False. The electrical charge of an unstimulated neuron is negative inside the membrane.

4. Clusters of rough endoplasmic reticulum, or sites of protein synthesis within a neuron, are called _____ _____.

Critical Thinking Questions

1. What is the importance of having a refractory period in the transmission of a nerve impulse?

Answers: 1. Efferent (or Motor) 2. F 3. T 5. Nissl bodies

Chapter 11

THE NERVOUS SYSTEM

Chapter Check-In

❑ Classifying the components of the peripheral and central nervous system

❑ Discovering the different regions of the brain and their functions

❑ Describing the circulation and production of cerebrospinal fluid

❑ Appreciating the organization and distribution of the spinal nerves

❑ Distinguishing between the sympathetic and parasympathetic components of the autonomic nervous system

The nervous system integrates and monitors the countless actions occurring simultaneously throughout the entire human body. Therefore, every task, no matter how menial, accomplished by a person is a direct result of the components of the nervous system. These actions can be under voluntary control, like touching a computer key, or can occur without your direct knowledge, like digesting food, releasing enzymes from the pancreas, or other unconscious acts.

It is difficult to understand all the complexities of the nervous system because the field of neuroscience has rapidly evolved over the past 20 years. Moreover, answers to new questions are being found almost daily. However, a thorough knowledge of the individual components of the nervous system and their functions will lead to a better understanding of how the human body works and facilitate the acquisition of knowledge in the future.

Nervous System Organization

The **nervous system** consists of two parts, shown in Figure 11-1:

■ The **central nervous system (CNS)** consists of the brain and spinal cord.

- The **peripheral nervous system (PNS)** consists of nerves outside the CNS.

Figure 11-1 Two parts of the nervous system.

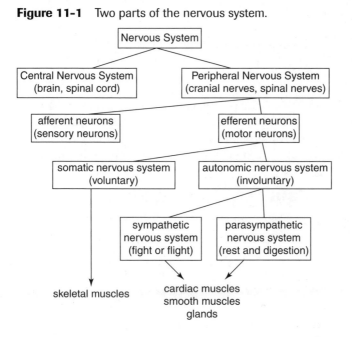

Nerves of the PNS are classified in three ways. First, PNS nerves are classified by how they are connected to the CNS. Cranial nerves originate from or terminate in the brain, while spinal nerves originate from or terminate at the spinal cord.

Second, nerves of the PNS are classified by the direction of nerve propagation. **Sensory (afferent) neurons** transmit impulses from skin and other sensory organs or from various places within the body to the CNS. **Motor (efferent) neurons** transmit impulses from the CNS to effectors (muscles or glands).

Third, motor neurons are further classified according to the effectors they target. The **somatic nervous system (SNS)** directs the contraction of skeletal muscles. The **autonomic nervous system (ANS)** controls the activities of organs, glands, and various involuntary muscles, such as cardiac and smooth muscles.

The autonomic nervous system has two divisions:

- The **sympathetic nervous system** is involved in the stimulation of activities that prepare the body for action, such as increasing the heart rate, increasing the release of sugar from the liver into the blood, and other activities generally considered as fight-or-flight responses (responses that serve to fight off or retreat from danger).

- The **parasympathetic nervous system** activates tranquil functions, such as stimulating the secretion of saliva or digestive enzymes into the stomach and small intestine.

Generally, both sympathetic and parasympathetic systems target the same organs, but often work antagonistically. For example, the sympathetic system accelerates the heartbeat, while the parasympathetic slows the heartbeat. Each system is stimulated as is appropriate to maintain homeostasis.

Nervous System Terminology

The following terms are commonly used in descriptions of nervous system features.

- A nerve fiber is an axon or dendrite. A nerve is a bundle of nerve fibers in the PNS. A nerve tract is a bundle of nerve fibers in the CNS.

- White matter consists of myelinated axons of neurons in the CNS.

- Gray matter consists of unmyelinated portions of neurons (cell bodies, dendrites, and axon terminals), unmyelinated neurons, and neuroglia in the CNS.

- **Nuclei** are clusters of cell bodies in the CNS. **Ganglia** are clusters of cell bodies in the PNS (except the basal ganglia, which are more appropriately called basal nuclei).

- Vesicles are fluid-filled cavities in the brain that form during early development. The tissues that form the vesicles divide to become the various components of the brain.

- Ventricles are interconnected cavities in the mature brain that originate from the fluid-filled vesicles. Circulating fluid (cerebrospinal fluid) in the ventricles provides nourishment for nervous tissue.

- Peduncles are large tracts that emerge from certain regions of the brain. Their large size gives the appearance of supporting the structure from which they emerge (peduncle means "little foot").

The Brain

Three cavities, called the primary brain vesicles, form during the early embryonic development of the brain. These are the forebrain (prosencephalon), the midbrain (mesencephalon), and the hindbrain (rhombencephalon).

During subsequent development, the three primary brain vesicles develop into five secondary brain vesicles. The names of these vesicles and the major adult structures that develop from the vesicles follow (see Table 11-1):

■ The telencephalon generates the cerebrum (which contains the cerebral cortex, white matter, and basal ganglia).

■ The diencephalon generates the thalamus, hypothalamus, and pineal gland.

■ The mesencephalon generates the midbrain portion of the brain stem.

■ The metencephalon generates the pons portion of the brain stem and the cerebellum.

■ The myelencephalon generates the medulla oblongata portion of the brain stem.

Table 11-1   The Vesicles and Their Components

Primary Vesicles	Secondary Vesicles	Adult Structure	Important Components or Features
prosencephalon (forebrain)	telencephalon	cerebrum (cerebral hemispheres)	cerebral cortex (gray matter): motor areas, sensory areas, association areas
prosencephalon (forebrain)	telencephalon	cerebrum (cerebral hemispheres)	cerebral white matter: association fibers, commisural fibers, projection fibers
prosencephalon (forebrain)	telencephalon	cerebrum (cerebral hemispheres)	basal ganglia (gray matter): caudate nucleus & amygdala, putamen, globus pallidus
prosencephalon	diencephalon	diencephalon	thalamus: relays sensory information

(continued)

Table 11-1 *(continued)*

Primary Vesicles	Secondary Vesicles	Adult Structure	Important Components or Features
prosencephalon (forebrain)	diencephalon	diencephalon	hypothalamus: maintains body homeostasis
prosencephalon (forebrain)	diencephalon	diencephalon	mammillary bodies: relays sensations of smells to cerebrum
prosencephalon (forebrain)	diencephalon	diencephalon	optic chiasma: crossover of optic nerves
prosencephalon (forebrain)	diencephalon	diencephalon	infundibulum: stalk of pituitary gland
prosencephalon (forebrain)	diencephalon	diencephalon	pituitary gland: source of hormones
prosencephalon forebrain)	diencephalon	diencephalon	epithalamus: pineal gland
mesencephalon (midbrain)	mesencephalon	brain stem	midbrain: cerebral peduncles, sup. cerebellar peduncles, corpora quadrigemina, superior colliculi
rhombencephalon (hindbrain)	metencephalon	brain stem	pons: middle cerebellar peduncles, pneumotaxic area, apneustic area
rhombencephalon (hindbrain)	metencephalon	cerebellum	sup. cerebellar peduncles, middle cerebellar peduncles, inferior cerebellar peduncles
rhombencephalon (hindbrain)	myelencephalon	brain stem	medulla oblongata: pyramids, cardiovascular center, respiratory center

A second method for classifying brain regions is by their organization in the adult brain. The following four divisions are recognized (see Figure 11-2).

Figure 11-2 The four divisions of the adult brain.

Lobes of the Cerebrum

Sagittal Section of the Brain

Ventricles of the Brain

■ The **cerebrum** consists of two cerebral hemispheres connected by a bundle of nerve fibers, the corpus callosum. The largest and most visible part of the brain, the cerebrum, appears as folded ridges and grooves, called convolutions. The following terms are used to describe the convolutions:

> A gyrus (plural, gyri) is an elevated ridge among the convolutions.
>
> A sulcus (plural, sulci) is a shallow groove among the convolutions.
>
> A fissure is a deep groove among the convolutions.

The deeper fissures divide the cerebrum into five lobes (most named after bordering skull bones)—the frontal lobe, the parietal love, the temporal lobe, the occipital lobe, and the insula. All but the insula are visible from the outside surface of the brain.

A cross section of the cerebrum shows three distinct layers of nervous tissue:

> The cerebral cortex is a thin outer layer of gray matter. Such activities as speech, evaluation of stimuli, conscious thinking, and control of skeletal muscles occur here. These activities are grouped into motor areas, sensory areas, and association areas.
>
> The cerebral white matter underlies the cerebral cortex. It contains mostly myelinated axons that connect cerebral hemispheres (association fibers), connect gyri within hemispheres (commissural fibers), or connect the cerebrum to the spinal cord (projection fibers). The corpus callosum is a major assemblage of association fibers that forms a nerve tract that connects the two cerebral hemispheres.
>
> Basal ganglia (basal nuclei) are several pockets of gray matter located deep inside the cerebral white matter. The major regions in the basal ganglia—the caudate nuclei, the putamen, and the globus pallidus—are involved in relaying and modifying nerve impulses passing from the cerebral cortex to the spinal cord. Arm swinging while walking, for example, is controlled here.

■ The **diencephalon** connects the cerebrum to the brain stem. It consists of the following major regions:

> The thalamus is a relay station for sensory nerve impulses traveling from the spinal cord to the cerebrum. Some nerve impulses are sorted and grouped here before being transmitted to the cerebrum. Certain sensations, such as pain, pressure, and temperature, are evaluated here also.

The **epithalamus** contains the pineal gland. The pineal gland secretes **melatonin,** a hormone that helps regulate the biological clock (sleep-wake cycles).

The hypothalamus regulates numerous important body activities. It controls the autonomic nervous system and regulates emotion, behavior, hunger, thirst, body temperature, and the biological clock. It also produces two hormones (ADH and oxytocin) and various releasing hormones that control hormone production in the anterior pituitary gland.

The following structures are either included or associated with the hypothalamus.

The mammillary bodies relay sensations of smell.

The infundibulum connects the pituitary gland to the hypothalamus.

The optic chiasma passes between the hypothalamus and the pituitary gland. Here, portions of the optic nerve from each eye cross over to the cerebral hemisphere on the opposite side of the brain.

■ The brain stem connects the diencephalon to the spinal cord. The brain stem resembles the spinal cord in that both consist of white matter fiber tracts surrounding a core of gray matter. The brain stem consists of the following four regions, all of which provide connections between various parts of the brain and between the brain and the spinal cord. (Some prominent structures are listed in Table 11-2 and illustrated in Figure 11-3).

The midbrain is the uppermost part of the brain stem.

The pons is the bulging region in the middle of the brain stem.

The medulla oblongata (medulla) is the lower portion of the brain stem that merges with the spinal cord at the foramen magnum.

The reticular formation consists of small clusters of gray matter interspersed within the white matter of the brain stem and certain regions of the spinal cord, diencephalon, and cerebellum. The reticular activation system (RAS), one component of the reticular formation, is responsible for maintaining wakefulness and alertness and for filtering out unimportant sensory information. Other components of the reticular formation are responsible for maintaining muscle tone and regulating visceral motor muscles.

Figure 11-3 Prominent structures of the brain stem.

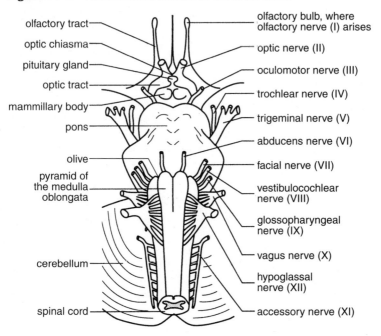

- The cerebellum consists of a central region, the vermis, and two wing-like lobes, the cerebellar hemispheres. Like that of the cerebrum, the surface of the cerebellum is convoluted, but the gyri, called folia, are parallel and give a pleated appearance. The cerebellum evaluates and coordinates motor movements by comparing actual skeletal movements to the movement that was intended.

The **limbic system** is a network of neurons that extends over a wide range of areas of the brain. The limbic system imposes an emotional aspect to behaviors, experiences, and memories. Emotions such as pleasure, fear, anger, sorrow, and affection are imparted to events and experiences. The limbic system accomplishes this by a system of fiber tracts (white matter) and gray matter that pervades the diencephalon and encircles the inside border of the cerebrum. The following components are included:

- The hippocampus (located in the cerebral hemisphere)
- The denate gyrus (located in cerebral hemisphere)
- The amygdala (amygdaloid body) (an almond-shaped body associated with the caudate nucleus of the basal ganglia)

- The mammillary bodies (in the hypothalamus)

- The anterior thalamic nuclei (in the thalamus)

- The fornix (a bundle of fiber tracts that links components of the limbic system)

The Ventricles and Cerebrospinal Fluid

There are four cavities in the brain, called ventricles. The ventricles are filled with cerebrospinal fluid (CSF), which provides the following functions:

- Absorbs physical shocks to the brain

- Distributes nutritive materials to and removes wastes from nervous tissue

- Provides a chemically stable environment

There are four ventricles:

- Each of two lateral ventricles occupies a cerebral hemisphere.

- The third ventricle is connected by a passage (interventricular foramen) to each of the two lateral ventricles.

- The fourth ventricle connects to the third ventricle (via the cerebral aqueduct) and to the central canal of the spinal cord (a narrow, central tube extending the length of the spinal cord). Additional openings in the fourth ventricle also allow CSF to flow into the subarachnoid space.

A network of capillaries called the choroids plexuses projects into each ventricle. Ependymal cells (a type of neuroglial cell) surround these capillaries. Blood plasma entering the ependymal cells from the capillaries is filtered as it passes into the ventricle, forming CSF. Any material passing from the capillaries to the ventricles of the brain must do so through the ependymal cells because tight junctions linking these cells prevent the passage of plasma between them. Thus, the ependymal cells maintain a blood-CSF barrier controlling the composition of the CSF.

The CSF circulates from the lateral ventricles (where most of the CSF is produced) to the third and then fourth ventricles. From the fourth ventricle, most of the CSF passes into the subarachnoid space, a space within the linings (meninges) of the brain, although some CSF also passes into the central canal of the spinal cord. The CSF returns to the blood through the arachnoid villi located in the dural sinuses of the meninges.

The Meninges

The meninges (singular, meninx) are protective coverings of the brain (cranial meninges) and spinal cord (spinal meninges). They consist of three layers of membranous connective tissue:

- The dura mater is the tough outer layer lying just inside the skull and vertebrae. Some characteristics follow:

   In the brain, there are channels within the dura mater, the dural sinuses, which contain venous blood returning from the brain to the jugular veins.

   In the spinal cord, the dura mater is often referred to as the dural sheath. A fat-filled space between the dura mater and the vertebrae, the epidural space, acts as a protective cushion to the spinal cord.

- The arachnoid (arachnoid mater) is the middle meninx. Projections from the arachnoid, called arachnoid villi, protrude through the dura mater into the dural sinuses. The arachnoid villi transport the CSF from the subarachnoid space to the dural sinuses. Two cavities border the arachnoid:

   The subdural space occurs outside the arachnoid (between the arachnoid and the dura mater).

   The subarachnoid space lies inside the arachnoid. This space contains blood vessels and circulates CSF. The fine threads of tissue that spread across this space resemble the web of a spider and give the arachnoid layer its name (arachnid means spider).

- The pia mater is the innermost meninx layer. It tightly covers the brain (following its convolutions) and spinal cord and carries blood vessels that provide nourishment to these nervous tissues.

The Blood-Brain Barrier

Cells in the brain require a very stable environment to ensure controlled and selective stimulation of neurons. As a result, only certain materials are allowed to pass from blood vessels to the brain. Substances such as O_2, glucose, H_2O, CO_2, essential amino aids, and most lipid-soluble substances enter the brain readily. Other substances, such as creatine and urea (wastes transported in the blood), most ions (Na^+, K^+, Cl^-), proteins, and certain toxins either have limited access or are totally blocked from entering the brain. Unfortunately, most antibiotic drugs are equally blocked from entering, while other substances such as caffeine, alcohol, nicotine, and heroin readily enter the brain (because of their lipid solubility). This blood-brain barrier is established by the following:

■ Brain capillaries are less permeable than other capillaries because of tight junctions between the endothelial cells in the capillary walls.

■ The basal lamina (secreted by the endothelial cells) that surrounds the brain capillaries decreases capillary permeability. This layer is usually absent in capillaries found elsewhere.

■ Processes from astrocytes (a type of neuroglial cell) cover brain capillaries and are believed to influence capillary permeability in some way.

Cranial Nerves

Cranial nerves are nerves of the PNS that originate from or terminate in the brain. There are 12 pairs of cranial nerves, all of which pass through foramina of the skull. Cranial nerves are either sensory nerves (containing only or predominately sensory fibers) or mixed nerves (containing both sensory and motor fibers). Characteristics of the cranial nerves, which are numbered from anterior to posterior as they attach to the brain, are summarized in Table 11-2 and illustrated in Figure 11-3, earlier in this chapter.

Table 11-2   Characteristics of Cranial Nerves

Cranial Nerve	Nerve Type	Major Functions
I Olfactory	sensory	smell
II Optic	sensory	vision
III Oculomotor	primarily motor	eyeball & eyelid movement; lens shape
IV Trochlear	primarily motor	eyeball movement; proprioception
V Trigeminal: ophthalmic branch	sensory	sensations of touch & pain from facial skin, nose, mouth, teeth, & tongue; proprioception motor control of chewing
V Trigeminal: maxillary branch	sensory	sensations of touch & pain from facial skin, nose, mouth, teeth, & tongue; proprioception motor control of chewing
V Trigeminal: mandibular branch	mixed	sensations of touch & pain from facial skin, nose, mouth, teeth, & tongue; proprioception motor control of chewing

(continued)

Table 11-2 *(continued)*

Cranial Nerve	Nerve Type	Major Functions
VI Abducens	primarily motor	eyeball movement; proprioception
VII Facial	mixed	movement of facial muscles; tear & saliva secretion; sense of taste & proprioception
VIII Vestibulocochlear: cochlear branch	sensory	hearing
VIII Vestibulocochlear: vestibular branch	sensory	sense of equilibrium
IX Glossophayrngeal	mixed	sensations of taste, touch & pain from tongue & pharynx; chemoreceptors (that monitor O_2 & CO_2), blood pressure receptors; movement of tongue & swallowing; secretion of saliva
X Vagus	mixed	parasympathetic sensation & motor control of smooth muscles associated with heart, lungs, viscera; secretion of digestive enzymes
XI Accessory	primarily motor	head movement; swallowing; proprioception
XII Hypoglossal	primarily motor	tongue movement, speech & swallowing; proprioception

The Spinal Cord

The spinal cord has two functions:

■ *Transmission of nerve impulses.* Neurons in the white matter of the spinal cord transmit sensory signals from peripheral regions to the brain and motor signals from the brain to peripheral regions.

■ *Spinal reflexes.* Neurons in the gray matter of the spinal cord integrate incoming sensory information and respond with motor impulses that control muscles (skeletal, smooth, or cardiac) or glands.

The **spinal cord** is an extension of the brain stem that begins at the foramen magnum and continues down through the vertebral canal to the first lumbar vertebra (L_1). Here, the spinal cord comes to a tapering point, the conus medullaris. The spinal cord is held in position at its inferior end by the filum terminale, an extension of the pia mater that attaches to the coccyx. Along its length, the spinal cord is held within the vertebral canal by denticulate ligaments, lateral extensions of the pia mater that attach to the dural sheath.

The following external features on the spinal cord (see Figure 11-4).

- Spinal nerves emerge in pairs, one from each side of the spinal cord along its length.

- The cervical enlargement is a widening in the upper part of the spinal cord (C_4 to T_1). Nerves that extend into the upper limbs originate or terminate here.

- The lumbar enlargement is a widening in the lower part of the spinal cord (T_9 to T_{12}). Nerves that extend into the lower limbs originate or terminate here.

- The anterior median fissure and the posterior median sulcus are two grooves that run the length of the spinal cord on its anterior and posterior surfaces, respectively.

- The cauda equina are nerves that attach to the end of the spinal cord and continue to run downward before turning laterally to other parts of the body.

Figure 11-4   External features of the spinal cord.

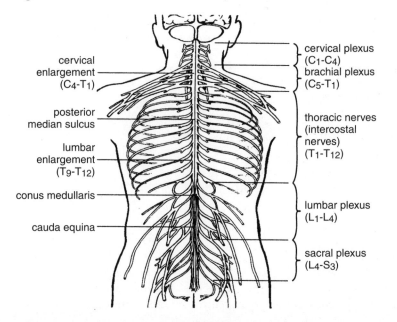

A cross section of the spinal cord reveals the following features, shown in Figure 11-5:

Figure 11-5   A cross-section of the spinal cord.

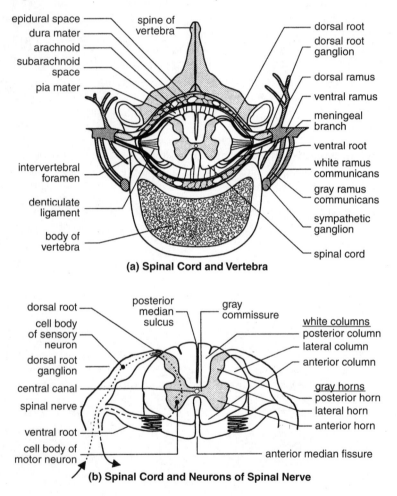

(a) Spinal Cord and Vertebra

(b) Spinal Cord and Neurons of Spinal Nerve

- Roots are branches of the spinal nerve that connect to the spinal cord. Two major roots form:

A ventral root (anterior or motor root) is the branch of the nerve that enters the ventral side of the spinal cord. Ventral roots contain motor nerve axons, transmitting nerve impulses from the spinal cord to skeletal muscles.

A dorsal root (posterior or sensory root) is the branch of a nerve that enters the dorsal side of the spinal cord. Dorsal roots contain sensory nerve fibers, transmitting nerve impulses from peripheral regions to the spinal cord.

A dorsal root ganglion is a cluster of cell bodies of a sensory nerve. It is located on the dorsal root.

■ Gray matter appears in the center of the spinal cord in the form of the letter H (or a pair of butterfly wings) when viewed in cross section.

The gray commissure is the cross-bra of the H.

The anterior (ventral) horns are gray matter areas at the front of each side of the H. Cell bodies of motor neurons that stimulate skeletal muscles are located here.

The posterior (dorsal) horns are gray matter areas at the rear of each side of the H. These horns contain mostly interneurons that synapse with sensory neurons.

The lateral horns are small projections of gray matter at the sides of H. These horns are present only in the thoracic and lumbar regions of the spinal cord. They contain cell bodies of motor neurons in the sympathetic branch of the autonomic nervous system.

The central canal is a small hole in the center of the H cross-bar. It contains CSF and runs the length of the spinal cord and connects with the fourth ventricle of the brain.

■ White columns (funiculi) refer to six areas of the white matter, three on each side of the H. They are the anterior (ventral) columns, the posterior (dorsal) columns, and the lateral columns.

■ Fasciculi are bundles of nerve tracts within white columns containing neurons with common functions or destinations.

Ascending (sensory) tracts transmit sensory information from various parts of the body to the brain.

Descending (motor) tracts transmit nerve impulses from the brain to muscles and glands.

Spinal Nerves

There are 31 pairs of spinal nerves (62 total). The following discussion traces a spinal nerve as it emerges from the spinal column (also refer to Figure 11-5):

■ A spinal nerve emerges at two points from the spinal cord, the ventral and dorsal roots.

■ The ventral and dorsal roots merge to form the whole spinal nerve.

■ The spinal nerve emerges from the spinal column through an opening (intervertebral foramen) between adjacent vertebrae. This is true for all spinal nerves except for the first spinal nerve (pair), which emerges between the occipital bone and the atlas (the first vertebra).

■ Outside the vertebral column, the nerve divides into the following branches:

> The dorsal ramus contains nerves that serve the dorsal portions of the trunk.
>
> The ventral ramus contains nerves that serve the remaining ventral parts of the trunk and the upper and lower limbs.
>
> The meningeal branch reenters the vertebral column and serves the meninges and blood vessels within.
>
> The rami communicantes contain autonomic nerves that serve visceral functions.

■ Some ventral rami merge with adjacent ventral rami to form a plexus, a network of interconnecting nerves. Nerves emerging from a plexus contain fibers from various spinal nerves, which are now carried together to some target location.

An area of the skin that receives sensory stimuli that pass through a single spinal nerve is called a dermatome. Dermatomes are illustrated on a human figure with lines that mark the boundaries of the area where each spinal nerve receives stimuli.

Reflexes

A reflex is a rapid, involuntary response to a stimulus. A reflex arc is the pathway traveled by the nerve impulses during a reflex. Most reflexes are spinal reflexes with pathways that traverse only the spinal cord. During a spinal reflex, information may be transmitted to the brain, but it is the spinal cord, and not the brain, that is responsible for the integration of sensory information and a response transmitted to motor neurons. Some reflexes are cranial reflexes with pathways through cranial nerves and the brain stem.

A reflex arc involves the following components, shown in Figure 11-6:

- The **receptor** is the part of the neuron (usually a dendrite) that detects a stimulus.

- The sensory neuron transmits the impulse to the spinal cord.

- The integration center involves one synapse (monosynaptic reflex arc) or two or more synapses (polysynaptic reflex arc) in the gray matter of the spinal cord. In polysynaptic reflex arcs, one or more interneurons in the gray matter constitute the integration center.

- A motor neuron transmits a nerve impulse from the spinal cord to a peripheral region.

- An **effector** is a muscle or gland that receives the impulse form the motor neuron. In somatic reflexes, the effector is skeletal muscle. In autonomic (visceral) reflexes, the effector is smooth or cardiac muscle, or a gland.

Figure 11-6 A reflex arc.

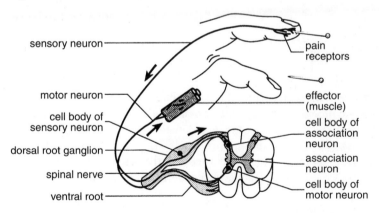

Some examples of reflexes follow:

- A stretch reflex is a monosynaptic reflex that is a response to a muscle that has been stretched (the knee jerk reflex is an example). When receptors in muscles, called muscle spindles, detect changes in muscle length, they stimulate, through a reflex arc, the contraction of a muscle. Stretch reflexes help maintain posture by stimulating muscles to regain normal body position.

■ A flexor (withdrawal) reflex is a polysynaptic reflex that causes a limb to be withdrawn when it encounters pain (refer to Figure 11-6).

The Autonomic Nervous System

The peripheral nervous system consists of the somatic nervous system (SNS) and the autonomic nervous system (ANS). The SNS consists of motor neurons that stimulate skeletal muscles. In contrast, the ANS consists of motor neurons that control smooth muscles, cardiac muscles, and glands. In addition, the ANS monitors visceral organs and blood vessels with sensory neurons, which provide input information for the CNS.

The ANS is further divided into the sympathetic nervous system and the parasympathetic nervous system. Both of these systems can stimulate and inhibit effectors. However, the two systems work in opposition—where one system stimulates an organ, the other inhibits. Working in this fashion, each system prepares the body for a different kind of situation, as follows.

■ The sympathetic nervous system prepares the body for situations requiring alertness or strength or situations that arouse fear, anger, excitement, or embarrassment ("fight-or-flight" situations). In these kinds of situations, the sympathetic nervous system stimulates cardiac muscles to increase the heart rate, causes dilation of the bronchioles of the lungs (increasing oxygen intake), and causes dilation of blood vessels that supply the heart and skeletal muscles (increasing blood supply). The adrenal medulla is stimulated to release epinephrine (adrenalin) and norepinephrine (noradrenalin), which in turn increases the metabolic rate of cells and stimulate the liver to release glucose into the blood. Sweat glands are stimulated to produce sweat. In addition, the sympathetic nervous system reduces the activity of various "tranquil" body functions, such as digestion and kidney functioning.

■ The parasympathetic nervous system is active during periods of digestion and rest. It stimulates the production of digestive enzymes and stimulates the processes of digestion, urination, and defecation. It reduces blood pressure and heart and respiratory rates and conserves energy through relaxation and rest.

In the SNS, a single motor neuron connects the CNS to its target skeletal muscle. In the ANS, the connection between the CNS and its effector consists of two neurons—the preganglionic neuron and the postganglionic neuron. The synapse between these two neurons lies outside the CNS, in an autonomic ganglion. The axon (preganglionic axon) of a preganglionic neuron enters the ganglion and forms a synapse with the dendrites of the postganglionic neuron emerges from the ganglion and travels to the target organ (see Figure 11-7). There are three kinds of autonomic ganglia:

■ The sympathetic trunk, or chain, contains sympathetic ganglia called paravertebral ganglia. There are two trunks, one on either side of the vertebral column along its entire length. Each trunk consists of ganglia connected by fibers, like a string of beads.

■ The prevertebral (collateral) ganglia also consist of sympathetic ganglia. Preganglionic sympathetic fibers that pass through the sympathetic trunk (without forming a synapse with a postganglionic neuron) synapse here. Prevertebral ganglia lie near the large abdominal arteries, which the preganglionic fibers target.

■ Terminal (intramural) ganglia receive parasympathetic fibers. These ganglia occur near or within the target organ of the respective postganglionic fiber.

Figure 11-7 The target organs of the different nervous systems.

	central nervous system	peripheral nervous system		target organs
somatic nervous system		myelin sheath	ACh	skeletal muscle
sympathetic nervous system		ACh paravertebral or prevertebral ganglion	NE	smooth muscle
para-sympathetic nervous system		myelin sheath terminal ganglion	ACh ACh	glands
				cardiac muscle
ACh = acetylocholine NE = norepinephrine		pre-ganglionic axon	ganglion	post-ganglionic axon

A comparison of the sympathetic and parasympathetic pathways follows (see Figure 11-8):

■ *Sympathetic nervous system.* Cell bodies of the preganglionic neurons occur in the lateral horns of gray matter of the 12 thoracic and first 2 lumbar segments of the spinal cord. (For this reason, the sympathetic system is also called the thoracolumbar division.) Preganglionic fibers leave the spinal cord within spinal nerves through the ventral roots (together with the PNS motor neurons). The preganglionic fibers then branch away from the nerve through white rami (white rami communicantes) that connect with the sympathetic trunk. White rami are white because they contain myelinated fibers. A preganglionic fiber that enters the trunk may synapse in the first ganglion it enters, travel up or down the trunk to synapse with another ganglion, or pass through the trunk and synapse outside the trunk. Postganglionic fibers that originate in ganglia within the sympathetic trunk leave the trunk through gray rami (gray rami communicantes) and return to the spinal nerve, which is followed until it reaches its target organ. Gray rami are gray because they contain unmyelinated fibers.

■ *Parasympathetic nervous system.* Cell bodies of the preganglionic neurons occur in the gray matter of sacral segments S_2-S_4 and in the brain stem (with motor neurons of their associated cranial nerves III, VII, IX, and X). (For this reason, the parasympathetic system is also called the craniosacral division, and the fibers arising from this division are called the cranial outflow or the sacral outflow, depending upon their origin.) Preganglionic fibers of the cranial outflow accompany the PNS motor neurons of cranial nerves and have terminal ganglia that lie near the target organ. Preganglionic fibers of the sacral outflow accompany the PNS motor neurons of spinal nerves. These nerves emerge through the ventral roots of the spinal cord and have terminal ganglia that lie near the target organ.

Figure 11-8 A comparison of the sympathetic and parasympathetic pathways.

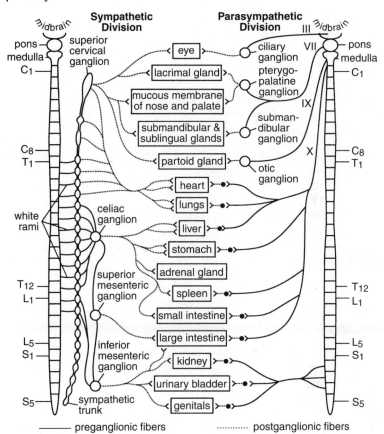

Chapter Checkout

Q&A

1. The _____ nervous system controls the activities of voluntary movement.

2. True or False. The sympathetic nervous system can simultaneously cause certain blood vessels to contract while others are stimulated to dilate.

3. Which of the following statement(s) is/are *true* concerning the diencephalon?

   a. forms the rostral most projection of the brainstem
   b. contains the superior colliculus
   c. primarily composed of the hypothalamus and thalamus
   d. derived from the prosencephalon
   e. located inferior to the cerebellum

4. True or False. The dorsal root carries afferent nerve fibers.

5. The _____ _____ is the tapered point of the spinal cord at the level of the second lumbar vertebra.

Answers: 1. efferent (or motor) 2. T (only the SNS regulates blood flow) 3. a, c, and d 4. T 5. Conus medullaris

Chapter 12

THE SENSORY SYSTEM

Chapter Check-In

❑ Finding out how sensory receptors are classified

❑ Understanding the different types of sensory stimuli that the body detects

❑ Naming the components of the eye and its accessory organs that aid in vision

❑ Describing the major regions of the ear and the process of hearing

❑ Appreciating the other special senses, such as taste, smell, and equilibrium

One of the characteristics of a living organism is its ability to respond to stimuli. The human sensory system is highly evolved and processes thousands of incoming messages simultaneously. This complexity allows you to be aware of your surroundings and take appropriate actions.

Sensory Receptors

Sensory receptors are dendrites of sensory neurons specialized for receiving specific kinds of stimuli. Sensory receptors are classified by three methods:

■ *Classification by receptor complexity.*

Free nerve endings are dendrites whose terminal ends have little or no physical specialization.

Encapsulated nerve endings are dendrites whose terminal ends are enclosed in a capsule of connective tissue.

Sense organs (such as the eyes and ears) consist of sensory neurons with receptors for the special senses (vision, hearing, smell, taste, and equilibrium) together with connective, epithelial, or other tissues.

■ *Classification by location.*

Exteroceptors occur at or near the surface of the skin and are sensitive to stimuli occurring outside or on the surface of the body. These receptors include those for tactile sensations, such as touch, pain, and temperature, as well as those for vision, hearing, smell, and taste.

Interoceptors (visceroceptors) respond to stimuli occurring in the body from visceral organs and blood vessels. These receptors are the sensory neurons associated with the autonomic nervous system.

Proprioceptors respond to stimuli occurring in skeletal muscles, tendons, ligaments, and joints. These receptors collect information concerning body position and the physical conditions of these locations.

■ *Classification by type of stimulus detected.*

Mechanoreceptors respond to physical force such as pressure (touch or blood pressure) and stretch.

Photoreceptors respond to light.

Thermoreceptors respond to temperature changes.

Chemoreceptors respond to dissolved chemicals during sensations of taste and smell and to changes in internal body chemistry such as variations of O_2, CO_2, or H+ in the blood.

The Somatic Senses

The somatic (general) senses collect information about cutaneous sensations (tactile sensations on the surface of the skin) and proprioceptive sensations. The following stimuli are detected:

■ Tactile stimuli are detected by mechanoreceptors and produce sensations of touch and pressure:

Merkel discs are receptors with free nerve endings that detect surface pressure (light touch). They are located at the bottom of the epidermis.

Root hair plexuses are receptors with free nerve endings that surround hair follicles and detect hair movement.

Corpuscles of touch (Meissner's corpuscles) are receptors with encapsulated nerve endings located in the dermal paillae (near the surface) of the skin that detect surface pressure (light touch).

Pacinian corpuscles are encapsulated nerve receptors that detect deep pressure and are located in the subcutaneous layer (below the skin).

- Thermal stimuli are detected by free nerve ending thermoreceptors sensitive to heat or cold.

- Pain stimuli are detected by free nerve ending nociceptors.

- Proprioceptive stimuli are detected by the following receptors:

    Muscle spindles are mechanoreceptors located in skeletal muscles. They consist of specialized skeletal muscle fibers enclosed in a spindle-shaped capsule made of connective tissue.

    Golgi tendon organs are mechanoreceptors located at junctions of tendons and muscles.

    Joint kinesthetic receptors are mechanoreceptors located in synovial joints.

Vision

The eye is supported by the following accessory organs:

- The eyebrows shade the eyes and help keep perspiration that accumulated on the forehead from running into the eyes.

- The eyelids (palpebrae) lubricate, protect, and shade the eyeballs. Contraction of the levator palpebrae superioris muscle raises the upper eyelid. Each eyelid is supported internally by a layer of connective tissue, the tarsal plate. Tarsal (Meibomian) glands embedded in the tarsal plate produce secretions that prevent the upper and lower eyelids from sticking together. The inner lining of the eyelid, the conjunctiva, is a mucous membrane that produces secretions that lubricate the eyeball. The conjunctiva continues beyond the eyelid, folding back to cover the white of the eye.

- The eyelashes, on the borders of the eyelids, help protect the eyeball. Nerve endings at the base of the hairs initiate reflex action that closes the eyelids when the eyelashes are disturbed.

- The lacrimal apparatus produces and drains tears. Tears (lacrimal fluid) are produced by the lacrimal glands, which lie above each eye (toward the outer side). In each eye, tears flow across the eyeball and enter two openings (lacrimal puncta) into lacrimal canals that lead to the lacrimal sac. From here, the tears drain through the nasolacrimal duct into the nasal cavity. Tears contain antibodies and lysozyme (a bacteria-destroying enzyme).

- Six extrinsic eye muscles provide fine motor control for the eyeballs. These are the lateral, medial, superior, and inferior rectus muscles and the inferior and superior oblique muscles.

The eyeball is a hollow sphere whose wall consists of three tunics (layers), shown in Figure 12-1:

- The outer fibrous tunic consists of avascular connective tissue. The forward 1/6 portion of this tunic is the cornea, a transparent layer of collagen fibers that forms a window for entering light. The remainder of the fibrous tunic is the sclera. Consisting of tough connective tissue, the sclera maintains the shape of the eyeball and provides for the attachment of the eye muscles. The visible forward portion of the sclera is the white of the eye.

    The **iris** is the colored portion of the eye that opens and closes to control the size of its circular opening, the **pupil.** The size of the pupil regulates the amount of light entering the eye and helps bring objects into focus.

    The ciliary body lies between the iris and the choroids (the remainder of the vascular tunic). The ciliary processes that extend from the ciliary body secrete aqueous humor, the fluid that fills the forward chamber of the eye. The suspensory ligament between the ciliary processes and the lens hold the lens in place, while ciliary muscles (in the ciliary body) that pull on the suspensory ligament control the shape of the lends to focus images.

    The choroid connects with the ciliary body at a jagged boundary, the and forms the remaining portion (5/6) of the vascular tunic. The choroid is dark brown, absorbing light and reducing reflection within the chamber of the eyeball that would otherwise blur images. The highly vascularized choroid provides nutrients to surrounding tissues, including the avascularized fibrous tunic.

Figure 12-1 Details of the eye and the retina.

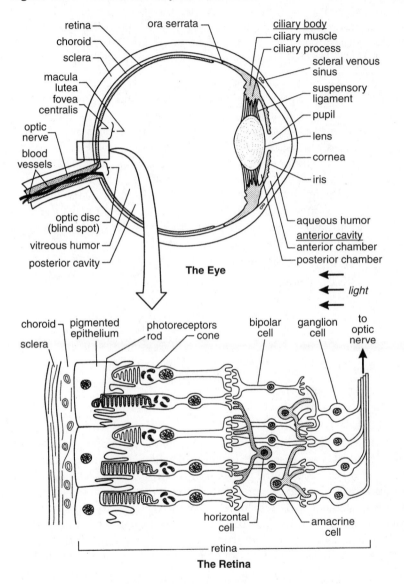

The Eye

The Retina

■ The middle vascular tunic (uvea) consists of three highly vascularized (as the name implies), pigmented parts:

■ The inner nervous tunic is the **retina.** The retina consists of an outer pigmented epithelium covered by nervous tissue (the neural layer) on the inside. The dark color of the pigmented epithelium absorbs light (as with the choroid) and stores vitamin A used by photoreceptor cells in the neural layer. There are two kinds of photoreceptors in the retina:

> **Cones** are photoreceptor cells that respond to bright light and color. They transmit sharp images. The concentration of cones is low at the sides of the retina and increases as the cones approach the macula lutea, an oval region in the center of the rear portion of the retina. The center of the macula lutea, the fovea centralis, contains only cones; other retinal cells are absent, exposing the cones directly to incoming light. The high concentration of cones and direct exposure to light make the fovea centralis the site on the retina that provides the highest visual acuity. As a result, images that are viewed directly are focused upon the fovea centralis.

> **Rods** are photoreceptor cells that are more sensitive to light and more numerous than cones. As a result, rods provide vision in dim light. They are also more capable of detecting movement. However, rods cannot detect color, and dimly lit objects appear gray. Because the concentration of rods increases in areas farther away from the macula lutea, detecting a moving or dimly lit object can be more effectively achieved by looking slightly away from the object.

Within the nervous tunic, photoreceptor cells (rods and cones) form synapses with other nerve cells (refer to Figure 12-1). When stimulated by light, rods and cones pass graded potentials to bipolar cells, which, in turn, pass graded potentials to the ganglion cells. The graded potentials may be modified by horizontal cells and amacrine cells that link adjacent photoreceptors or ganglion cells, respectively. Action potentials are ultimately generated by ganglion cells. The axons of all the ganglion cells gather at the optic disc and exit the nervous tunic through the optic disc as the optic nerve. The optic disc is a blind spot because photoreceptors are absent here.

The lens of the eye consists of tightly packed cells arranged in successive layers (as in an onion) and filled with transparent proteins called crystallins. The lens divides the interior of the eyeball into two cavities:

■ The anterior cavity, the area in front of the lens, is subdivided by the iris and ciliary body into the anterior chamber and the posterior chamber. Capillaries in the ciliary body produce a clear fluid, the aqueous humor, which flows into the posterior chamber, through the pupil, and into the anterior chamber. The aqueous humor then drains into veins through a channel (scleral venous sinus, or canal of Schlemm) that encircles the eye where the cornea and sclera join. The aqueous humor, which is continuously replaced, provides pressure to maintain the shape of the forward portion of the eye and supplies O_2 and nutrients to the avascular lens and cornea.

■ The posterior cavity, the area behind the lens, is filled with a clear gel, the vitreous humor. The vitreous humor, which is produced during embryonic development and is not replaced, holds the lens and retina in position and maintains the shape of the eye.

The **process of sight** involves converting light energy to chemical energy. Feature of the process follow:

■ The outer segments of rods and cones contain numerous folds that increase the surface area exposed to light.

■ Photopigments (visual pigments) in the outer segments respond to light by changing their chemical structure. Each visual photopigment consists of two parts—retinal (a vitamin A derivative) and opsin (a glycoprotein). There are four different kinds of photopigments because the opsin in each has a slightly different structure, enabling each photopigment to absorb a different range of light wavelengths (different colors).

■ There are three kinds of cones, each possessing a different photopigment, and each sensitive to a different range of light wavelengths. A mixed stimulation of the three different cones (called red, green, and blue for their optimally absorbed wavelengths) provides for the perception of varied colors.

■ There is only one kind of rod, with a single kind of photopigment, rhodopsin.

■ When a photopigment absorbs light, retinal changes shape, causing it to separate from opsin. Because the product is colorless, the process is called bleaching.

■ Unlike most other neurons, photoreceptors continually secrete a neurotransmitter when in the resting (unstimulated), polarized condition. When a photoreceptor is stimulated, the freed opsin becomes chemically active and initiates a series of chemical reactions that close the Na⁺ channels in the plasma membrane, Because the Na⁺/K⁺ pump continues to pump Na⁺ out of the cell, the plasma membrane becomes hyperpolarized. Hyperpolarization stops the normal secretion of the neurotransmitter, which in turn stimulates a graded potential in the bipolar cells.

■ Photopigments are regenerated when opsin is enzymatically reattached to retinal. In bright light, the very light-sensitive rhodospin in rods cannot be regenerated as fast as it is broken down, so most of the rhodospin remains inactive. As a result, the less light-sensitive pigments in cones are active in normal daylight. When eyes move from bright conditions to dark conditions, the photopigments in cones are too insensitive to detect light, and the rhodospin in rods is still bleached from its exposure to bright light. Vision returns as rhodospin is regenerated, a process called dark adaptation. When eyes move from dark to bright conditions, very light-sensitive rods are suddenly overwhelmed with stimulation, producing the sensation of glare. Light adaptation occurs as the rhodospin in rods is completely bleached and the less light-sensitive cones resume activity.

When light reflected from an object enters the eye, the following processes occur:

■ *Light refraction.* When light rays pass from one substance to another substance of different density, the rays bend, or refract. The amount of bending depends upon the angle of incidence of the light ray and the degree to which the densities of the two substances differ. When distant objects are sighted, the normal curvature of the lens appropriately compensates for the refraction due to the differences in densities among the aqueous humor, the lens, and the vitreous humor.

■ *Lens accommodation.* Light rays from near objects enter the eyeball at more divergent angles than rays from distant objects. Thus, when near objects are sighted, muscles pull on the lens to increase its curvature so that the more divergent rays of the close object are properly refracted upon the retina.

■ *Pupil constriction.* One function of the pupil is to regulate the amount of light that enters the posterior cavity so that the retina receives the appropriate amount of stimulation. In addition, when near objects are sighted, the pupil constricts (accommodation papillary reflex) to block the most divergent light rays that cannot be brought into focus by the accommodation of the lens. Reading under low levels of light may be difficult because the pupils are dilated to allow all available light to enter rather than constricted to improve focusing. Increasing the amount of light improves the ability to read because the surplus light permits the pupils to constrict and the lens to focus a more narrow beam of light rays.

■ *Eyeball convergence.* Both eyes point in the same direction when viewing a distant object. When near objects are sighted, the eyes must be directed medially to simultaneously view the object, a process called convergence.

Nerve impulses generated by visual stimuli travel along the axons of ganglion cells within the two optic nerves. Before entering the brain, axons representing the medial portions of the visual fields of each eye cross over at the optic chiasma. After the crossover, the axons, now forming the optic tract, enter the thalamus. Processed visual stimuli are then carried to visual areas of the occipital lobes of both cerebral hemispheres by nerve pathways called the optic radiations. Because of the partial crossover at the optic chiasma, each cerebral hemisphere receives the lateral portion of the visual field of the eye on the same side of the body as the cerebral hemisphere, but the medial portion of the visual field of the eye on the opposite side of the body. In addition, because of the action of the lens, the image that forms on the retina and that is sent to the brain is inverted and reversed right to left. The brain, however, interprets all of this seemingly disparate visual information into a coherent perception of the real world.

Hearing

The organ of hearing, the ear, consists of three major regions, shown in Figure 12-2:

■ The outer (external) ear consists of the auricle (pinna), a flap of elastic cartilage that protrudes from the head, and the external auditory canal (meatus), a tube that enters the temporal bone. The canal is lined with ceruminous glands that secrete cerumen (earwax), a sticky substance that traps dirt and other foreign objects. The eardrum (tympanic membrane), at the internal end of the external auditory canal, vibrates in response to incident sound waves.

Figure 12-2 The three major regions of the ear are the middle ear, and the inner ear.

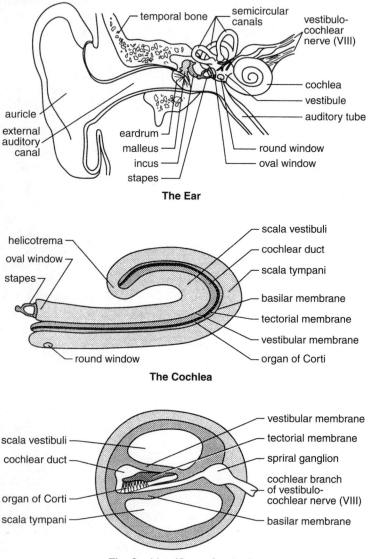

The Ear

The Cochlea

The Cochlea (Cross Section)

- The middle ear (tympanic cavity) is an air-filled cavity within the temporal bone. It contains three small bones, the auditory ossicles. The bones, called the malleus (hammer), incus (anvil), and stapes (stirrup), act as a lever system that amplifies and transfers vibrations of the eardrum to the inner ear. The malleus at one end connects to the eardrum, while the stapes, at the other end, attaches with ligaments to the oval window, a small, membrane-covered opening into the inner ear. Synovial joints connect the incus, the center bone of the auditory ossicles, to the malleus and stapes on each side. A second membrane-covered opening to the inner ear, the round window (secondary tympanic membrane), lies just below the oval window. A third opening leads to the auditory (Eustachian) tube, which connects the middle ear to the upper throat. The auditory tube allows pressure differences between the middle and outer ear to equalize, thus reducing tension on the eardrum. Two muscles in the middle ear, the tensor tympani and the stapedius, connect to the malleus and stapes, respectively. Contraction of these two muscles restricts the movement of the eardrum and auditory ossicles, reducing damage that may occur when they are exposed to excessive vibration from loud noises.

- The inner (internal) ear, also called the labyrinth, is a system of double-walled canals. The canals consist of an outer bony (osseous) labyrinth that encloses an inner membranous labyrinth. Perilymph fills the space between the two labyrinths, and endolymph fills the inner labyrinth. This double-layer labyrinth structure is found throughout the following inner ear structures.

Three semicircular canals contain receptor cells for determining angular movements of the head. This information is used for establishing equilibrium.

The cochlea is a coiled canal that contains receptor cells that respond to vibrations transferred from the middle ear. The interior of the cochlea is divided into three regions, or scalas—the scala vestibuli, the scala tympani, and the cochlear duct (scala media). The scalas are tubular channels that follow the coiled curvature of the cochlea. At the middle ear, the scala vestibuli and the scala tympani connect to the oval and round windows, respectively. At the other end of the cochlea, a region called the helioctrema, these two scalas join, allowing free movement of the perilymph within. The third scala, the cochlear duct, is separated from the scala vestibuli and the scala tympani by the vestibular membrane and the basilar membrane, respectively. The cochlear duct is filled with endolymph and internally lined with the organ of Corti. The organ of Corti contains numerous hair cells. The bases of the hair cells are attached to the basilar membrane, while hairlike microvilli called stereocilia project upwards into an overlying gel, the tectorial membrane. The stereocilia are receptors for vibrations that are produced when the underlying basal membrane moves relative to the overlying tectorial membrane.

The vestibule lies between the semicircular canals and the cochlea. It contains two bulblike sacs, the saccule and utricle, whose membranes are continuous with those of the cochlea and semicircular canals, respectively. The saccule and utricle contain receptors that help maintain equilibrium.

The process of hearing occurs as follows:

1. Sound waves, funneled into the outer ear by the auricle, cause the eardrum to vibrate.

2. Vibrations of the eardrum are amplified and transferred by the auditory ossicles of the oval window.

3. Vibrations on the oval window produce pressure waves in the perilymph of the scala vestibuli and the scala tympani. These vibrations are transferred to the basilar membrane.

4. Vibrations of the basilar membrane move the hair cells of the organ of Corti. The stereocilia of the hair cells bend when they move against the tectorial membrane. The bending generates a graded potential in the hair cell, which causes the release of a neurotransmitter at its base. The neurotransmitter, in turn, generates an action potential in dendrites of the cochlear nerve. Cell bodies of the cochlear nerve assemble in the spiral ganglia, and its axons merge with the vestibulocochlear nerve.

5. Pressure waves in the perilymph of the scala tympani cause the round window to bulge into the middle ear. This allows vibrational movements of the perilymph (and indirectly the endolymph) that, as an incompressible fluid, would not otherwise be able to vibrate within the surrounding rigid temporal bone.

Equilibrium

Equilibrium is maintained in response to two kinds of motion:

■ Static equilibrium maintains the position of the head in response to linear movements of the body, such as starting to walk or stopping.

■ Dynamic equilibrium maintains the position of the head in response to rotational motion of the body, such as rocking (as in a boat) or turning.

The perception of equilibrium occurs in the vestibular apparatus, which consists of the vestibule and the semicircular canals. Motion in these two structures is detected as follows:

■ The vestibule is the primary detector of changes in static equilibrium. A sensory receptor called a macula is located in the walls of the saccule and utricle, the two bulblike sacs of the vestibule. A macula contains numerous receptor cells called hair cells, from which numerous stereocilia (long microvilli) and a single kinocilium (a true cilium) extend into a glycoprotein gel, the otolithic membrane. Calcium carbonate crystals called otoliths pervade the otolithic membrane, increasing its density and thus responsiveness to changes in motion. Changes in linear motion cause the otolithic membrane to move forward and backward in the utricle or up and down in the saccule. The movement of the otolithic membrane causes similar movements in the embedded stereocilia of the hair cells, which in turn initiate graded potentials.

■ The semicircular canals are the primary detector of changes in dynamic equilibrium. The three canals, individually called the anterior, posterior, and lateral canals, are arranged at right angles to one another. The expanded base of each canal, called an ampulla, contains a sensory receptor, or crista ampullaris. Like the maculae of the vestibule, each crista contains numerous hair cells whose stereocilia and kinocilium protrude into a gelatinous matrix, the cupula (which is analogous to the otolithic membranes of the maculae). Changes in rotational motion cause the cupula and the embedded stereocilia to move, which stimulates the hair cell to generate a graded potential.

Graded potentials in the hair cells of the maculae and cristae result in changes in the amounts of neurotransmitter secreted. In response to these changes, action potentials are generated in the fibers of the vestibular nerve, which subsequently joins the vestibulocochlear nerve. From here, the nerve impulses travel to the pons and the cerebellum.

Smell

The sense of smell, or **olfactory sense,** occurs in olfactory epithelium that occupies a small area in the roof of the nasal cavity. The olfactory receptor cells are bipolar neurons whose dendrites have terminal knobs with hairlike cilia protruding beyond the epithelial surface. The cilia, or olfactory hairs, initiate an action potential when they react with a molecule from an inhaled vapor. However, molecules of the vapor must first dissolve in the mucus that covers the cilia before they can be detected. The action potential is transmitted along the axons of the olfactory receptor cells (which form the olfactory nerves) to the olfactory bulbs, where they synapse with sensory neurons of the olfactory tract.

Other cells of the olfactory epithelium include columnar supporting cells and basal cells. The basal cells continually divide to produce new olfactory receptor cells that, because of their short life, need regular replacement. The replacement of olfactory receptor cells is unusual because most other nerve cells cannot be replaced.

The mucus that lines the olfactory epithelium is produced by olfactory (Bowman's) glands that occupy the connective tissue above the olfactory epithelium.

Taste

The sense of taste, or **gustatory sense,** occurs in the taste buds. Located primarily on the tongue, taste buds reside in papillae, the bumps on the tongue that give it a rough texture. The taste bud consists of supporting cells, basal cells, and gustatory (taste) receptor cells arranged in the shape of a glove with an opening, or taste pore, to the outside located at the top. A long microvilli, or gustatory hair, from each gustatory receptor cell within the taste bud projects through the taste pore. Gustatory hairs generate action potentials when stimulated by chemicals that are dissolved in the saliva.

Basal cells are actively dividing epithelial cells. The daughter cells of basal cells develop into supporting cells, which subsequently mature into gustatory receptor cells. Because they are easily damaged by the activities that occur in the mouth, gustatory receptor cells are short-lived and replaced about every ten days.

An individual gustatory receptor cell responds to only four taste sensations: sweet, bitter, sour, and salty. All other tastes arise from a mixture of these four tastes in combination with olfactory sensations associated with the substance tasted. Taste buds on certain areas of the tongue seem to specialize for certain tastes. For example, the sensation of sweetness is best detected at the front of the tongue, while bitterness is best detected at the back of the tongue.

Chapter Checkout

Q&A

1. True or False. The cochlear duct (or scala media) is filled with perilymph.
2. Which of the following statement(s) is/are true regarding vision?
   a. Photopigments in the outer segments alter their structure in response to light
   b. Three kinds of rods are within the retina.
   c. Cones provide vision in dim light.
   d. During accommodation, the pupil constricts when near objects are sighting.
4. True or False. Proprioception is realizing the position in which the body currently resides.
5. _____ equilibrium maintains the position of the head in response to rotational motion of the body.

Critical Thinking Questions

1. Which cranial nerves are involved with each special sense?

Answers: 1. F 2. a, d 3. T 4. Dynamic

Chapter 13

THE ENDOCRINE SYSTEM

Chapter Check-In

❑ Classifying the four different types of hormones

❑ Describing how hormones affect their target organs and help regulate homeostasis within the body

❑ Listing the hormones within the body and their functions

It seems like a farfetched notion that a small chemical can enter the bloodstream and cause an action at a very distant location in the body. Yet this scenario occurs everyday. The ability to maintain homeostasis and respond to stimuli is largely due to hormones secreted within your body. Without hormones, you could not grow, maintain a constant temperature, produce offspring, or perform the basic actions that are essential for life.

Hormones

The **endocrine system** produces hormones that are instrumental in maintaining homeostasis and regulating reproduction and development. A hormone is a chemical messenger produced by a cell that effects specific change in the cellular activity of other cells (target cells). Unlike exocrine glands (which produce substances such as saliva, milk, stomach acid, and digestive enzymes), endocrine glands do not secrete substances into ducts (tubes). Instead, endocrine glands secrete their hormones directly into the surrounding extracellular space. The hormones then diffuse into nearby capillaries and are transported throughout the body in the blood.

The endocrine and nervous systems often work toward the same goal—both influence other cells with chemicals (hormones and neurotransmitters). However, they attain their goals differently. Neurotransmitters act immediately (within milliseconds) on adjacent muscle, gland, or other nervous cells, and their effect is short-lived. In contrast, hormones take longer to produce their intended effect (seconds to days), may affect any cell, nearby or distant, and produce effects that last as long as they remain in the blood (up to several hours).

Hormones can be chemically classified into four groups:

■ Amino acid-derived hormones are modified amino acids.

■ Polypeptide and protein hormones are chains of amino acids of less than or more than about 100 amino acids, respectively. Some protein hormones are actually glycoproteins, containing glucose or other carbohydrate groups.

■ Steroid hormones are lipids that are synthesized from cholesterol. Steroids are characterized by four interlocking carbohydrate rings.

■ Eicosanoids are lipids that are synthesized from the fatty acid chains of phospholipids found in plasma membrane.

Mechanisms of hormone action

Hormones circulating in the blood diffuse into the interstitial fluids surrounding the cell. Cells with specific receptors for a hormone respond with an action that is appropriate for the cell. Because of the specificity of hormone and target cell, the effects produced by a single hormone may vary among different kinds of target cells.

Hormones activate target cells by one of two methods, depending upon the chemical nature of the hormone:

■ Lipid-soluble hormones (steroid hormones and hormones of the thyroid gland) diffuse through the cell membranes of target cells. The lipid-soluble hormone then binds to a receptor protein that, in turn, activates a DNA segment that turns on specific genes. The proteins produced as result of the transcription of the genes and subsequent translation of mRNA act as enzymes that regulate specific physiological cell activity.

■ Water-soluble hormones (polypeptide, protein, and most amino acid hormones) bind to a receptor protein on the plasma membrane of the cell. The receptor protein, in turn, stimulates the production of one of the following second messengers:

■ Cyclic AMP (cAMP) is produced when the receptor protein activates another membrane-bound protein called a G protein. The G protein activates adenylate cyclase, the enzyme that catalyzes the production of cAMP from ATP. Cyclic AMP then triggers an enzyme that generates specific cellular changes.

- Inositol triphosphate (IP_3) is produced from membrane phospholipids. IP_3, in turn, triggers the release of Ca^{2+} from the endoplasmic reticulum, which then activates enzymes that generate cellular changes.

Control of hormone production

Endocrine glands release hormones in response to one (or more) of the following stimuli:

- Hormones form other endocrine glands.

- Chemical characteristics of the blood (other than hormones).

- Neural stimulation.

Most hormone production is regulated by a **negative feedback** system. The nervous system and certain endocrine tissues monitor various internal conditions of the body. If action is necessary to maintain homeostasis, hormones are released, either directly by an endocrine gland or indirectly though the action of the hypothalamus of the brain, which stimulates other endocrine glands to release hormones. The hormones activate target cells, which initiate physiological changes that adjust body conditions. When normal conditions have been restored, the corrective action—the production of hormones—is discontinued. Thus, in negative feedback, when the original (abnormal) condition has been repaired, or negated, corrective actions decrease (or are discontinued). For example, the amount of glucose in the blood regulates the secretion of insulin and glucagons through negative feedback.

The production of some hormones is regulated by **positive feedback.** In such a system, hormones cause a condition to intensify (rather than decrease). As the condition intensifies, hormone production increases. Such positive feedback is uncommon but does occur during childbirth (hormone levels build with increasingly intense labor contractions) and lactation (where hormone levels increase in response to nursing, which causes milk production to increase).

The Hypothalamus and Pituitary Glands

The hypothalamus makes up the lower region of the diencephalons and lies just above the brain stem. The pituitary gland (hypophysis) is attached to the bottom of the hypothalamus by a slender stalk called the infundibulum. The pituitary gland consists of two major regions—the anterior pituitary gland (anterior lobe or adenohypophysis) and the posterior pituitary gland (posterior lobe or neurohypophysis).

The hypothalamus oversees many internal body conditions. It receives nervous stimuli from receptors throughout the body and monitors chemical and physical characteristics of the blood, including temperature, blood pressure, and nutrient, hormone, and water content. When deviations from homeostasis occur or when certain developmental changes are required, the hypothalamus stimulates cellular activity in various parts of the body by directing the release of hormones from the anterior and posterior pituitary glands. The hypothalamus communicates directives to these glands by one of the following two pathways:

- Communication between the hypothalamus and the anterior pituitary occurs through chemicals (releasing hormones and inhibiting hormones) that are produced by the hypothalamus and delivered to the anterior pituitary through blood vessels. The releasing and inhibiting hormones are produced by specialized neurons of the hypothalamus called neurosecretory cells. The hormones are released into a capillary network (primary plexus) and transported through veins (hypophyseal portal veins) to a second capillary network (secondary plexus) that supplies the anterior pituitary. The hormones then diffuse from the secondary plexus into the anterior pituitary, where they initiate the production of specific hormones by the anterior pituitary. The releasing and inhibiting hormones secreted by the hypothalamus and the hormones produced in response by the anterior pituitary are listed in Table 13-1. Many of the hormones produced by the anterior pituitary are tropic hormones (tropins), hormones that stimulate other endocrine glands to secrete their hormones.

- Communication between the hypothalamus and the posterior pituitary occurs through neurosecretory cells that span the short distance between the hypothalamus and the posterior pituitary. Hormones produced by the cell bodies of the neurosecretory cells are packaged in vesicles and transported through the axon and stored in the axon terminals that lie in the posterior pituitary. When the neurosecretory cells are stimulated, the action potential generated triggers the release of the stored hormones from the axon terminals to a capillary network within the posterior pituitary. Two hormones, oxytocin and antidiuretic hormone (ADH), are produced and released in this way. Their functions are summarized in Table 13-1.

Table 13-1   Hormone Functions

Source	Hormone (H), Releasing Hormone (RH), Or Inhibiting Hormone (IH)	Chemical Form*	Target	Action
Hypothalamus				
GHRH	growth hormone RH	PP	anterior pituitary	inhibits release of hGH
GHIH	growth hormone IH (somatostatin)	PP	anterior pituitary	stimulates release of hGH
TRH	thyrotropin RH	PP	anterior pituitary	stimulates release of TSH and hGH
GnRH	gonadotropin RH	PP	anterior pituitary	stimulates release of LH and FSH
PRH	prolactin RH	PP	anterior pituitary	stimulates release of PRL
PIH	prolactin IH (dopanmine)	PP	anterior pituitary	inhibits release of PRL
CRH	corticotropin RH	PP	anterior pituitary	stimulates release of ACTH
Anterior pituitary (tropic hormones)				
TSH	thyroid stimulating H (thyrotropin)	GP	thyroid	stimulates secretion of T_3 and T_4
ACTH	adrenocortico-tropic hormone	PP	adrenal cortex	stimulates secretion of gluco-corticoids
FSH	follicle-stimulating hormone	GP	ovary, testes	regulates oogenesis & sperma-togenesis

Source	Hormone (H), Releasing Hormone (RH), Or Inhibiting Hormone (IH)	Chemical Form*	Target	Action
Anterior pituitary (tropic hormones)				
LH	luteinizing hormone	GP	ovary, testes	regulates oogenesis & spermatogenesis
Anterior pituitary (hormones)				
PRL	prolactin	PR	mammary glands	stimulates production of milk
hGH	human growth H (somatotropin)	PR	bone, muscle, various	stimulates growth
Posterior pituitary				
OT	oxytocin	PP	uterus, mammary glands	uterine contractions, release of milk
ADH	antidiuretic H (vasopressin)	PP	kidneys, sweat glands	increases water retention
Thyroid gland				
T_4	thyroxine	AA	most body cells	increases rate of cellular metabolism
T_3	triiodothyronine	AA	bone	increases rate of cellular metabolism
	calcitonin	PP	bone	decreases blood Ca^{2+}
Parathyroid gland				
PTH	parathyroid hormone	PP	bone, kidneys, intestine	increases blood Ca^{2+}

(continued)

·le 13-1 *(continued)*

Source	Hormone (H), Releasing Hormone (RH), Or Inhibiting Hormone (IH)	Chemical Form*	Target	Action
Adrenal medulla				
NE	epinephrine (adrenaline)	AA	blood vessels, liver, heart	increases blood sugar, constricts blood vessels flight response)
NE	norepinephrine (noradrenaline)	AA	blood vessels, liver, heart	increases blood sugar constricts blood vessels (fight or flight response)
Adrenal cortex				
	mineralocorticoids (e.g., aldosterone)	S	kidneys	increase reabsorption of Na^+, excretion of K^+
	glucocorticoids (e.g., cortisol)	S	most body cells	increase blood sugar
	androgens (e.g., DHEA)	S	general	stimulate onset of puberty, female sex drive
Pancreas				
	glucagon (secreted by alpha cells)	PP	liver	increases blood glucose
	insulin (secreted by beta cells)	PP	liver, muscle, adipose	decreases blood glucose
	somatostatin (secreted by delta cells)	PP	alpha & beta cells	inhibits insulin & glucagon release

Source	Hormone (H), Releasing Hormone (RH), Or Inhibiting Hormone (IH)	Chemical Form*	Target	Action
Pancreas				
	pancreatic polypeptide (from F cells)	PP	delta cells	inhibits somato-statin & pancreatic enzymes
Ovaries				
	estrogen	S	uterus, general	menstrual cycle, secondary sex charact-eristics
	progesterone	S	uterus	regulates menstrual cycle, pregnancy
	relaxin	PP	pelvis, cervix	dilates cervix & birth canal
	inhibin	PR	anterior pituitary	inhibits FSH re-lease
Testes				
	testosterone	S	testes, general	sperma-togenesis, secondary sex char-acteristics
	inhibin	PR	anterior pituitary	inhibits FSH release
Pineal				
	melatonin	AA	various	regulates biological clock

(continued)

Table 13-1 *(continued)*

Source	Hormone (H), Releasing Hormone (RH), Or Inhibiting Hormone (IH)	Chemical Form*	Target	Action
Kidney				
	erythropoietin	GP	bone marrow	increases blood cell production
	calcitriol (Vitamin D)	S	intestine	increases Ca^{2+} absorption
Placenta				
	estrogen	S	uterus	maintains pregnancy, mammary glands
	progesterone	S	uterus	maintains pregnancy, mammary glands
	hCG	GP	ovary	stimulates release of estrogen & progesterone
	hCS	PR	mammary glands	prepares mammary glands for lactation
Gastrointestinal tract				
	gastrin	PP	stomach	stimulates HCl release
GIP	gastrin inhibitory peptide	PP	stomach, pancreas	inhibits gastric juice release, increases insulin
	secretin	PP	pancreas, liver	stimulates release of enzymes & bile

Source	Hormone (H), Releasing Hormone (RH), Or Inhibiting Hormone (IH)	Chemical Form*	Target	Action
Gastrointestinal tract				
CCK	cholecystokinin	PP	pancreas, liver	stimulates release of enzymes & bile
	serotonin	AA	stomach	stimulates stomach muscle contraction
Heart				
ANP	atrial natriuretic peptide	PP	kidney, adrenal cortex	decreases blood pressure
Most cells				
PG	prostaglandins	E	all cells except red blood cells	various
LT	leukotrienes	E	all cells except red blood cells	various

Endocrine Organs and Tissues

A summary of the various endocrine organs, their hormones, and their functions is given in Table 13-1. Also listed are some organs whose major function is not the secretion of hormones, but which, nonetheless, contain some specialized cells that produce hormones. These organs include the heart, the gastrointestinal tract, the placenta, the kidneys, and the skin.

In addition, all cells (except red blood cells) secrete a class of hormones called eicosanoids. These hormones are paracrines, or local hormones, that primarily affect neighboring cells. Two groups of eicosanoids, the prostaglandins (PGs) and the leukotrienes (LTs), have a wide range of varying effects that depend upon the nature of the target cell. Eicosanoid activity, for example, may impact blood pressure, blood clotting, immune and inflammatory responses, reproductive processes, and the contraction of smooth muscles.

Antagonistic Hormones

Maintaining homeostasis often requires conditions to be limited to a narrow range. When conditions exceed the upper limit of homeostasis, specific action, usually the production of a hormone, is triggered. When conditions return to normal, hormone production is discontinued. If conditions exceed the lower limit of homeostasis, a different action, usually the production of a second hormone, is triggered. Hormones that act to return body conditions to within acceptable limits from opposite extremes are called **antagonistic hormones.**

The regulation of blood glucose concentration (through negative feedback) illustrates how the endocrine system maintains homeostasis by the action of antagonistic hormones. Bundles of cells in the pancreas called the islets of Langerhans contain two kinds of cells, alpha cells and beta cells. These cells control blood glucose concentration by producing the antagonistic hormones insulin and glucagon:

- Beta cells secrete insulin. When the concentration of blood glucose rises (after eating, for example), beta cells secret insulin into the blood. Insulin stimulates the liver and most other body cells to absorb glucose. Liver and muscle cells convert the glucose to glycogen (for short-term storage), and adipose cells convert the glucose to fat. In response, glucose concentration decreases in the blood, and insulin secretion discontinues (through negative feedback from declining levels of glucose).

- Alpha cells secrete glucagon. When the concentration of blood glucose drops (during exercise, for example), alpha cells secrete glucagon into the blood. Glucagon stimulates the liver to release glucose. The glucose in the liver originates from the breakdown of glycogon and the conversion of amino acids and fatty acids into glucose. When blood glucose levels return to normal, glucagon secretion discontinues (negative feedback).

Another example of antagonistic hormones occurs in the maintenance of Ca^{2+} concentration in the blood. Parathyroid hormone (PTH) from the parathyroid glands increases Ca^{2+} in the blood by increasing Ca^{2+} absorption in the intestines and reabsorption in the kidneys and stimulating Ca^{2+} release from bones. Calcitonin (CT) produces the opposite effect by inhibiting the breakdown of bone matrix and decreasing the release of calcium into the blood.

Chapter Checkout

Q&A

1. True or False. Lipid-soluble hormones bind to a receptor protein on the receptor membrane stimulating production of a second messenger.

2. All of the following are groups of hormones, *except*
   a. polypeptide
   b. steroid
   c. thyroxine
   d. amino acid derived

3. True or False. Calcitonin inhibits the breakdown of bone, decreasing the release of calcium into the blood.

4. Hormones that act to return the body to within acceptable limits from extreme conditions are called _____ hormones.

Critical Thinking Questions

1. Compare and contrast a positive and negative feedback system.

Answers: 1. F 2. c 3. T 4. antagonist

Chapter 14

THE CARDIOVASCULAR SYSTEM

Chapter Check-In

❑ Understanding the three main functions of the cardiovascular system

❑ Detailing the pathway of blood through the chambers of the heart

❑ Describing the cardiac cycle and cardiac output

❑ Knowing the three kinds of blood vessels and which direction blood flows through them

❑ Identifying the mechanisms that help regulate blood pressure

Knowing the functions of the cardiovascular system and the parts of the body that are part of it are critical in understanding the physiology of the human body. The cardiovascular system is the system that keeps life pumping through you with its complex pathways of veins, arteries, and capillaries. The heart, blood vessels, and blood help to transport vital nutrients throughout the body as well as remove metabolic waste. They help to protect the body and regulate body temperature.

The Functions

The **cardiovascular system** consists of the heart, blood vessels, and blood. This system has three main functions:

■ *Transport* of nutrients, oxygen, and hormones to cells throughout the body and removal of metabolic wastes (carbon dioxide, nitrogenous wastes, and heat).

■ *Protection* of the body by white blood cells, antibodies, and complement proteins that circulate in the blood and defend the body against foreign microbes and toxins. Clotting mechanisms are also present that protect the body from blood loss after injuries.

■ *Regulation* of body temperature, fluid pH, and water content of cells.

The Blood

The blood consists of cells and cell fragments, called **formed elements,** and water with dissolved molecules, called blood plasma (see Table 14-1).

Table 14-1   Formed Elements and Plasma of Blood

Constituent	*Characteristics/Functions*
Formed Elements (45%)	
Erythrocytes (red blood cells) (98–99%)	anucleate, contain hemoglobin; O_2 & CO_2 transport
Leukocytes (white blood cells) (0.1–0.3%):	
Neutrophils (60–70%)	granulocytes, polymorphonuclear; phagocytosis, wound healing
Eosinophils (2–4%)	granulocytes, bilobed nucleus; phagocytosis
Basophils (0.5–1%)	granulocytes, 2 to 5 lobed nucleus; release histamine
Lymphocytes (20–25%)	agranulocytes, circular nucleus, T cells, B cells; immune response, antibodies
Monocytes (3–8%)	agranulocytes, large kidney-shaped nucleus; phagocytotic macrophages
Thrombocytes (platelets) (1–2%)	anucleate, megakaryocyte fragments; blood clotting
Blood Plasma (55%)	
Water (90%)	
Plasma Proteins (8%):	
Albumin (54%)	maintain osmotic pressure between blood & tissues
Globulins (38%)	lipid and metal ion transporters, antibodies
Fibrinogen (7%)	clotting factor
Others (1%)	enzymes, hormones, clotting factors
Electrolytes:	Na^+, K^+, Ca^{2+}, Mg^{2+}, Cl^-, HCO_{3-}, SO_{42-}, HPO_{42-}

(continued

Table 14-1 *(continued)*

Constituent	Characteristics/Functions
Formed Elements (45%)	
Blood Plasma (55%)	
Gases:	O_2, CO_2, N_2
Nutrients: Glucose, other carbohydrates Amino acids Lipids Cholesterol	 sources of energy protein-building blocks fats, steroids, phospholipids component of plasma membranes & steroid hormones
Waste products: Urea Creatinine Uric acid Bilirubin	 from breakdown of proteins from breakdown of creatine phosphate (from muscles) from breakdown of nucleic acids from breakdown of hemoglobin
Hormones:	Various

Eyrthrocytes

Erythrocytes, or **red blood cells (RBCs),** transport oxygen (O_2) and carbon dioxide (CO_2) in the blood. Erythrocytes contain the protein hemoglobin to which both O_2 and CO_2 attach.

Mature erythrocytes lack a nucleus and most cellular organelles, thereby maximizing the cell's volume and thus its ability to carry hemoglobin and to transport O_2.

Erythrocytes are shaped like flattened donuts with a depressed center (rather than a donut hole). Their flattened shape maximizes surface area for the exchange of O_2 and CO_2 and allows flexibility that permits their passage through narrow capillaries.

Hemoglobin contains both a protein portion, called globin, and nonprotein heme groups. Globin consists of four polypeptide chains, each of which contains a heme group. The heme group is a red pigment that contains a single iron atom surrounded by a ring of nitrogen-containing carbon rings. One oxygen atom attaches to the iron of each heme group, allowing a single hemoglobin molecule to carry four oxygen atoms. Each erythrocyte contains about 250 million hemoglobin molecules.

Oxyhemoglobin (HbO_2) forms in the lungs when erythrocytes are exposed to oxygen as they pass through the lungs. Deoxyhemoglobin (Hb) forms when oxygen detaches form the iron and diffuses into surrounding tissues.

Carbaminohemoglobin ($HbCO_2$) forms when CO_2 attaches to amino acids of the globin part of the hemoglobin molecule. About 25 percent of the CO_2 transported from tissues to lungs is in this form.

Carbonic anhydrase, an enzyme in erythrocytes, converts CO_2 and H_2O in the blood plasma to H^+ and HCO_3. About 65% of the CO_2 collected from tissues travels in the blood plasma as HCO_3.

Because they lack cellular organelles and thus physiology to maintain themselves, erythrocytes survive for only about 120 days. Degenerated erythrocytes are broken down in the spleen and liver by macrophages (phagocytic white blood cells) as follows:

1. The globin and heme parts of the hemoglobin are separated. The globin is reduced to amino acids, which are returned to the blood plasma.

2. Iron is removed from the heme group and bound to the proteins ferritin and hemosiderin, which store the iron for later use (because unbound iron is toxic). Iron is also attached to transferrin, which enters the bloodstream. Transferrin may be picked up by muscles or liver cells, where it may be stored as ferritin or hemosiderin or picked up by bone marrow, where the iron is used to produce new erythrocytes.

3. The remainder of the heme group is broken down into bilirubin (a yellow-orange pigment), which enters the bloodstream and is picked up by the liver. Liver cells incorporate bilirubin into bile, which enters the small intestine during the digestion of fats. Bilirubin is then converted into urobilinogen by intestinal bacteria. Finally, most urobilinogen is converted to the brown pigment stercobilin, which is eliminated with the feces (and which gives feces its brown color). A small amount of urobilinogen is absorbed into the blood, converted to the yellow pigment urobilin, picked up by the kidneys, and eliminated with the urine (contributing to the yellow color of urine).

Leukocytes

Leukocytes, or **white blood cells (WBCs),** are cells that protect the body from foreign microbes and toxins. Although all leukocytes can be found in the bloodstream, some permanently leave the bloodstream to enter tissues where they encounter microbes or toxins, while other kinds of leukocytes readily move in and out of the bloodstream. Leukocytes are classified into two groups, granulocytes and agranulocytes, based upon the presence or absence of granules in the cytoplasm and the shape of the nucleus.

- Granulocytes contain numerous granules in the cytoplasm and have a nucleus that is irregularly shaped with lobes. Each of the three types of granulocytes is named after the bloodstains that its granules absorb.

- Neutrophils, the most numerous of granulocytes, have an S- or C-shaped nucleus with three to six lobes. Their granules, which are small and inconspicuous, poorly absorb both basic and acidic stains (neutral pH preference), producing a pale, lilac color. Because the shape of the nucleus is so variable, neutrophils are referred to as polymorphonuclear leukocytes (PMNs), or polys. Young neutrophils, with immature nuclei that are shaped like rods, are called band neutrophils. Neutrophils are the first leukocytes to arrive at a site of infection, responding (by chemotaxis) to chemicals released by damaged cells. The neutrophils, by phagocytosis, actively engulf bacteria, which are then destroyed by the various antibiotic proteins (such as defensins and lysozymes) contained within the granules. The neutrophils, usually destroyed in the process, contribute, together with other dead tissue, to the formation of pus.

- Eosinophils have a bilobed nucleus (two lobes connected by a narrow strand of chromatin). Their granules, which stain red with acid (eosin) dyes, contain digestive enzymes and are considered lysosomes. Eosinophils actively phagocytize complexes formed by the action of antibodies on antigens (foreign substances). Numbers of eosinophils increase during parasitic infection and allergic reactions.

- Basophils have a U- or S-shaped nucleus with two to five lobes connected by a narrow strand of chromatin. Their granules, which stain blue-purple with basic dyes, contain histamine, serotonin, and heparin. Basophils release histamine in response to tissue damage and to pathogen invasion (as part of the inflammatory response). Basophils resemble mast cells, cells similar in appearance and function to basophils, but found only in connective tissues.

- Agranulocytes, the second group of leukocytes, do not have visible granules in the cytoplasm and the nucleus is not lobed. There are two types of these leukocytes:

- Lymphocytes, often classified as small, medium, and large, have a roughly round nucleus surrounded by a small amount of blue-staining cytoplasm. Lymphocytes are the only leukocytes that return to the bloodstream, circulating among the bloodstream, tissue fluids, tissues, and lymph fluid. There are two major groups of lymphocytes, which vary based upon their role in an immune response. T lymphocytes (T cells), which mature in the thymus gland, attack aberrant cells (such as tumor cells, organ transplant cells, or cells infected by viruses). B lymphocytes (B cells), which mature in the bone marrow, respond to circulating antigens (such as toxins, viruses, or bacteria) by dividing to produce plasma cells, which in turn, produce antibodies.

- Monocytes have a large, kidney-shaped nucleus surrounded by ample blue-gray-staining cytoplasm. When monocytes leave the bloodstream and move into tissues, they enlarge and become macrophages, which engulf microbes and cellular debris.

Platelets

Platelets (thrombocytes) are fragments of huge cells called megakaryocytes. Platelets lack a nucleus and consists of cytoplasm (with few organelles) surrounded by a plasma membrane. Platelets adhere to damaged blood vessel walls and release enzymes that activate **hemostasis,** the stoppage of bleeding.

Plasma

Plasma is the straw-colored, liquid portion of the blood. It consists of the following:

- Water (90 percent).

- Proteins (8 percent). Albumin, the most common protein, is produced by the liver and serves to preserve osmotic pressure between blood and tissues. Other proteins include alpha and beta globulins (proteins that transport lipids and metal ions), gamma globulins (antibodies), fibrinogen and prothrombin (clotting proteins), and hormones.

- Waste products (urea, uric acid, creatinine, bilirubin, and others).

- Nutrients (absorbed from the digestive tract).

- Electrolytes (various ions such as sodium, calcium, chloride, and bicarbonate).

- Respiratory gases (O_2 and CO_2).

Serum is the liquid material remaining after blood-clotting proteins have been removed from plasma as a result of clotting.

Blood Formation

Hemopoiesis (hematoiesis) is the process that produces the formed elements of the blood. Hemopoiesis takes place in the red bone marrow found in the epiphyses of long bones (for example, the humerus and femur), flat bones (ribs and cranial bones), vertebrae, and the pelvis. Within the red bone marrow, hemopoietic stem cells (hemocytoblasts) divide to produce various "blast" cells. Each of these cells mature and becomes a particular formed element.

Erythropoiesis

Erythropoiesis, the process of making erythrocytes, begins with the formation of proerythroblasts from hemopoietic stem cells. Over three to five days, several stages of development follow as ribosomes proliferate and hemoglobin is synthesized. Finally, the nucleus is ejected, producing the depression in the center of the cell. Young erythrocytes, called reticulocytes, still containing some ribosomes and endoplasmic reticulum, pass into the bloodstream and develop into mature erythrocytes after another one or two days.

Erythroprotein

Erythroprotein (EPO), a hormone produced mostly by the kidneys, stimulates bone marrow to produce erythrocytes. When inadequate amounts of oxygen are delivered to body cells, a condition called hypoxia, the kidneys increase EPO secretion, which, in turn, stimulates an increase in erythrocyte production.

The average production rate of erythrocytes in healthy individuals is two million cells per second. Normal production requires adequate amounts of iron and vitamin B_{12} and folic acid.

Leukopoiesis

Leukopoiesis, the process of making leukocytes, is stimulated by various colony-stimulating factors (CSFs), hormones produced by mature white blood cells. The development of each kind of white blood cell begins with the division of themopoietic stem cells into one of the following "blast" cells.

■ Myeoblasts divide to form eosinophilic, neutrophilic, or basophilic myelocytes, which lead to the development of the three kinds of granulocytes.

■ Monoblasts lead to the development of monocytes.

■ Lymphoblasts lead to the development of lymphocytes.

Thrombopoiesis

Thrombopoiesis, the process of making platelets, begins with the formation of megakaryoblasts from hemopoietic stem cells. The megakaryoblasts divide without cytokinesis to become megakaryocytes, huge cells with a large, multilobed nucleus. The megakaryocytes then fragment into segments as the plasma membrane infolds into the cytoplasm.

Hemostasis

Hemostasis, the stoppage of bleeding, is accomplished through three steps:

1. A vascular spasm, a constriction of the damaged blood vessel, occurs at the site of injury. Vasoconstriction is initiated by the smooth muscle of the blood vessel in response to the injury and by nerve signals from pain receptors.

2. A platelet plug, consisting of a mass of linked platelets, fills the hole in the damaged blood vessel. Platelet plug formation follows these steps:

   a. *Platelet adhesion.* Platelets adhere to the exposed collagen fibers in the damaged blood vessel wall.

   b. *Platelet release.* Platelets release ADP (which attracts other platelets to the injury), serotonin (which stimulates vasoconstriction), and thromboxane A_2 (which attracts platelets and stimulates vasoconstriction). Cellular extensions from the platelets interconnect and form a loose mesh.

   c. *Platelet aggregation.* Additional platelets arrive at the site of the injury in response to the released ADP and expand the accumulation of platelets.

3. Coagulation (blood clotting) is a complex series of reactions that transform liquid blood into a gel (clot) that provides a secure patch

to the injured blood vessel. Thirteen coagulation factors (numbered I through XIII in order of their discovery) are involved. Most of these factors are proteins released into the blood by the liver. Factor III is Ca^{2+}. Vitamin K is required for the synthesis of some of these factors. The coagulation process can be described in three major steps.

a. *Formation of factor X and prothrombinase.* Prothrombinase (porthrombin activator) can form either intrinsically (inside the blood vessels) or extrinsically (outside the blood vessels). In the intrinsic pathway, the collagen of the damaged blood vessel initiates a cascade of reactions that activates factor X. In the extrinsic pathway, damaged tissues release thromboplastin (tissue factor, TF), which initiates a shorter and more rapid sequence of reactions to activate factor X. In both pathways, activated factor X combines with factor V (with Ca^{2+} present) to form prothrombinase.

b. *Prothrombin is converted to thrombin.* In this common pathway that follows both the intrinsic and extrinsic pathways, prothrombinase (with Ca^{2+}) converts prothrombin to thrombin.

c. *Fibrinogen is converted to fibrin.* The common pathway continues as thrombin (with Ca^{2+}) converts fibrinogen to fibrin. Fibrin forms long strands that bind the platelets together to form a dense web. Thrombin also activates factor XIII, which helps fibrin strands cohere to one another. The result is a clot.

Following its formation, a clot is further strengthened by a process called clot retraction. Platelets in the clot contract, pulling on the fibrin strands to which they are attached. The result is a more tightly sealed patch.

Fibrinolysis is the breakdown of the clot as the damaged blood vessel is repaired. During the formation of a clot, the plasma protein plasminogen is incorporated into the clot. The healthy endothelial tissue that replaces the damaged areas of the blood vessel secretes tissue plasminogen activator (t-PA), which converts plasminogen into its active form, plasmin (fibrinolysin). Plasmin, in turn, breaks down fibrin and leads to the dissolution of the clot.

Blood Groups

Various glycoproteins and lipoproteins are embedded in the surfaces of red blood cells. These proteins are inherited, and their structures may vary from one individual to another. If during a transfusion an individual receives blood containing RBCs with proteins that the individual does not carry, then these proteins may be recognized as antigens (foreign substances) by the immune system. If so, antibodies are produced that bind to the antigens and cause agglutination (clumping) and subsequent destruction of the foreign RBCs.

There are over 30 common groups of RBC proteins (referred to as antigens, isoantigens, or agglutinogens). Generally, each group is controlled by a single gene, and for each gene, two alleles, or forms, of the gene are inherited (one allele from each parent). Each blood group gene may have two or more different alleles in the population. Although not all blood group proteins stimulate the immune response, two important ones do.

- ■ ABO blood group. The gene responsible for this group has three alleles. One allele produces an "A" antigen, a second produces a "B" antigen, and a third produces no antigens ("O"). Because individuals inherit two alleles, individuals may be of the A blood type, inheriting two A alleles (AA) or an A and an O allele (AO), the B blood type (BB or BO), the AB blood type (AB), or the O blood type (OO). The immune response is activated when an individual receives a transfusion with blood carrying nonself antigens. For example, the immune system would respond if a person with A blood type (either AA or AO) receives blood of the B or AB blood types, but not of the O type. (The O type does not carry any foreign antigens.)

- ■ **Rh blood group.** This is a complex group defined by antigens produced by three different genes. Each gene has two (or rarely, three) alleles. Because of the close linkage of the genes (they are positioned close to one another on the same chromosome), the expression of the group can be evaluated as if it were a single gene with two alleles, an Rh^+ allele (producing the Rh antigen) and an Rh^- allele (producing no Rh antigen). Thus, individuals are either Rh^+ if they inherit one or two Rh^+ alleles or Rh^- if they inherit two Rh^- alleles.

Circulatory Pathways

Blood is confined to a closed system of blood vessels and to the four chambers of the heart (essentially dilated vessels). Blood travels away from the heart through **arteries,** which branch into smaller vessels, the arterioles. Arterioles branch further into the smallest vessels, the **capillaries.** Gas, nutrient, and waste exchange occurs across the capillary walls. The blood returns to the heart as capillaries merge to form venules, which further merge to form large **veins,** which connect to the heart. Blood circulates through the following two separate circuits.

- In the **pulmonary circulation,** deoxygenated blood travels from the right side of the heart to each of the two lungs. Within the lungs, O_2 enter and CO_2 leaves the capillaries by diffusion. Oxygenated blood returns from the lungs to the left side of the heart.

- In the **systemic circulation,** oxygenated blood travels from the left side of the heart to the various areas of the body. Gas, nutrient, and waste exchange occurs across the capillary walls into the interstitial fluids outside the capillaries and then into the surrounding cells. The deoxygenated blood returns to the right side of the heart.

The Heart

The heart is located in the mediastinum, the cavity between the lungs. The heart is tilted so that its pointed end, the apex, points downward toward the left hip, while the broad end, the base, faces upward toward the right shoulder. The heart is surrounded by the pericardium, a sac characterized by the following two layers:

- The outer fibrous pericardium anchors the heart to the surrounding structures.

- The inner serous pericardium consists of an outer parietal layer and an inner visceral layer. A thick layer of serous fluid, the pericardial fluid, lies between these two layers to provide a slippery surface for the movements of the heart.

The wall of the heart consists of three layers:

- The epicardium is the visceral layer of the serous pericardium.

- The myocardium is the muscular part of the heart that consists of contracting cardiac muscle and noncontracting Purkinje fibers that conduct nerve impulses.

- The endocardium is the thin, smooth, endothelial, inner lining of the heart, which is continuous with the inner lining of the blood vessels.

As blood travels through the heart, it enters a total of four chambers and passes through four valves. The two upper chambers, the right and left **atria,** are separated longitudinally by the **interatrial septum.** The two lower chambers, the right and left **ventricles,** are the pumping machines of the heart and are separated longitudinally by the **interventricular**

septum. A valve follows each chamber and prevents the blood from flowing backward into the chamber from which the blood originated.

Two prominent grooves are visible on the surface of the heart:

- The coronary sulcus (artioventricular groove) marks the junction of the atria and ventricles.

- The anterior interventricular sulcus and posterior interventricular sulcus mark the junction of the ventricles on the front and back of the heart, respectively.

The pathway of blood through the chambers and valves of the heart is described as follows (see Figure 14-1):

- The **right atrium,** located in the upper right side of the heart, and a small appendage, the right auricle, act as a temporary storage chamber so that blood will be readily available for the right ventricle. Deoxygenated blood from the systemic circulation enters the right atrium through three veins, the superior vena cava, the inferior vena cava, and the coronary sinus. During the interval when the ventricles are not contracting, blood passes down through the right atrioventricular (AV) valve into the next chamber, the right ventricle. The AV valve is also called the tricuspid valve because it consists of three flexible cusps (flaps).

- The **right ventricle** is the pumping chamber for the pulmonary circulation. The ventricle, with walls thicker and more muscular than those of the atrium, contracts and pumps deoxygenated blood through the three-cusped pulmonary semilunar valve and into a large artery, the pulmonary trunk. The pulmonary trunk immediately divides into two pulmonary arteries, which lead to the left and right lungs, respectively. The following events occur in the right ventricle.

   When the right ventricle contracts, the right AV valve closes and prevents blood from moving back into the right atrium. Small tendonlike cords, the chordae tendineae, are attached to papillary muscles at the opposite, bottom side of the ventricle. These cords limit the extent to which the AV valve can be forced closed, preventing it from being pushed through and into the atrium.

   When the right ventricle relaxes, the initial backflow of blood in the pulmonary artery closes the pulmonary semilunar valve and prevents the return of blood to the right ventricle.

■ The **left atrium** and its auricle appendage receive oxygenated blood from the lungs though four pulmonary veins (two from each lung). The left atrium, like the right atrium, is a holding chamber for blood in readiness for its flow into the left ventricle. When the ventricles relax, blood leaves the left atrium and passes through the left AV valve into the left ventricle. The left AV valve is also called the mitral or bicuspid valve, the only heart valve with two cusps.

■ The **left ventricle** is the pumping chamber for the systemic circulation. Because a greater blood pressure is required to pump blood through the much more extensive systemic circulation than through the pulmonary circulation, the left ventricle is larger and its walls are thicker than those of the right ventricle. When the left ventricle contracts, it pumps oxygenated blood through the aortic semilunar valve, into a large artery, the aorta, and throughout the body. The following events occur in the left ventricle, simultaneously and analogously with those of the right ventricle.

> When the left ventricle contracts, the left AV valve closes and prevents blood from moving back into the right atrium. As in the right AV valve, the chordae tendineae prevent overextension of the left AV valve.

> When the left ventricle relaxes, the initial backflow of blood in the aorta closes the aortic semilunar valve and prevents the return of blood to the left ventricle.

Two additional passageways are present in the fetal heart:

■ The foramen ovale is an opening across the interatrial septum. It allows blood to bypass the right ventricle and the pulmonary circuit, while the nonfunctional fetal lungs are still developing. The opening, which closes at birth, leaves a shallow depression called the fossa ovalis in the adult heart.

■ The ductus arteriosus is a connection between the pulmonary trunk and the aorta. Blood that enters the right ventricle is pumped out through the pulmonary trunk. Although some blood enters the pulmonary arteries (to provide oxygen and nutrients to the fetal lungs), most of the blood moves directly into the aorta through the ductus arteriosus.

Figure 14-1 The pathway of blood through the chambers and valves of the heart.

left pulmonary artery

right pulmonary artery

pulmonary trunk

superior vena cava

fossa ovalis

right atrium

right AV valve

pulmonary semilunar valve

right ventricle

inferior vena cava

chordae tendineae

papillary muscles

brachiocephalic artery

left common carotid artery

left subclavian artery

aortic arch

aorta

left pulmonary veins

right pulmonary veins

left atrium

left AV valve

aortic semilunar valve

left ventricle

interventricular septum

descending aorta

(a) Frontal Section of Heart

sinoatrial (SA) node

atrioventricular (AV) node

AV bundle (bundle of His)

right auricle

right atrium

right coronary artery

anterior cardiac veins

marginal artery

small cardiac vein

coronary sinus

posterior interventricular artery

middle cardiac vein

Purkinje fibers

right and left bundle branches

left coronary artery

left auricle

circumflex artery

great cardiac vein

anterior interventricular artery

(c) Conduction System of Heart

(b) Anterior View of Heart
(drawn as if transparent to show posterior blood vessels)

The **coronary circulation** consists of blood vessels that supply oxygen and nutrients to the tissues of the heart. Blood entering the chambers of the heart cannot provide this service because the endocardium is too thick for effective diffusion (and only the left side of the heart contains oxygenated blood). Instead, the following two arteries that arise from the aorta and encircle the heart in the artioventricular groove provide this function:

■ The left coronary artery has the following two branches: The anterior interventricular artery (left anterior descending, or LAD, artery) and the circumflex artery.

■ The right coronary artery has the following two branches: The posterior interventricular artery and the marginal artery.

Blood from the coronary circulation returns to the right atrium by way of an enlarged blood vessel, the coronary sinus. Three veins, the great cardiac vein, the middle cardiac vein, and the small cardiac vein, feed the coronary sinus.

Cardiac Conduction

Unlike skeletal muscle fibers (cells), which are independent of one another, cardiac muscle fibers (contractile muscle fibers) are linked by intercalated discs, areas where the plasma membranes intermesh. Within the intercalated discs, the adjacent cells are structurally connected by desmosomes, tight seals that weld the plasma membranes together, and electrically connected by gap junctions, ionic channels that allow the transmission of a depolarization event. As a result, the entire myocardium functions as a single unit with a single contraction of the atria followed by a single contraction of the ventricles.

Action potentials (electrical impulses) in the heart originate in specialized cardiac muscle cells called autorhythmic cells. These cells are self-excitable, able to generate an action potential without external stimulation by nerve cells. The autorhythmic cells serve as a pacemaker to initiate the cardiac cycle (pumping cycle of the heart) and provide a conduction system to coordinate the contraction of muscle cells throughout the heart. The autorhythmic cells are concentrated in the following areas (refer to Figure 14-1).

■ The sinoatrial (SA) node, located in the upper wall of the right atrium, initiates the cardiac cycle by generating an action potential that spreads through both atria through the gap junctions of the cardiac muscle fibers.

- The atrioventricular (AV) node, located near the lower region of the interatrial septum, receives the action potential generated by the SA node. A slight delay of the electrical transmission occurs here, allowing the atria to fully contract before the action potential is passed on to the ventricles.

- The atrioventricular (AV) bundle (bundle of His) receives the action potential from the AV node and transmits the impulse to the ventricles by way of the right and left bundle branches. Except for the AV bundle, which provides the only electrical connection, the atria are electrically insulated from the ventricles.

- The Purkinje fibers are large-diameter fibers that conduct the action potential from the interventricular septum, down to the apex, and then upward through the ventricles.

Cardiac Muscle Contraction

The sarcolemma (plasma membrane) of an unstimulated muscle cell is polarized—that is, the inside of the sarcolemma is negatively charged with respect to the outside. The unstimulated state of the muscle cell, called the resting potential, is created by the presence of large, negatively charged proteins and nucleic acids inside the cell. A balance between K^+ inside the cell and Na^+ outside the cell contributes to the polarization. During an action potential, the balance of Na^+ and K^+ is upset so that the cell becomes depolarized. The series of events that occurs during and following an action potential in contractile muscle fibers of the heart is similar to that in skeletal muscle. Here is a description of these events:

- Rapid depolarization occurs when fast-opening Na^+ channels in the sarcolemma open and allow an influx of Na^+ ions into the cardiac muscle cell. The Na^+ channels rapidly close.

- A plateau phase occurs during which Ca^{2+} enters the cytosol of the muscle cell. Ca^{2+} enters from the sarcoplasmic reticulum (endoplasmic reticulum) within the cell and also from outside the cell through slow-opening Ca^{2+} channels in the sarcolemma. Within the cell, Ca^{2+} binds to troponin, which, in turn, triggers the cross-bridge binding that leads to the sliding of actin filaments past myosin filaments. The sliding of the filaments produces cell contraction. At the same time that the Ca^{2+} channels open, K^+ channels, which normally leak small amounts of K^+ out of the cell, become more impermeable to K^+ leakage. The combined effects of the prolonged release of Ca^{2+} and the restricted leakage of K^+ lead to an extended depolarization that appears as a plateau when membrane potential is plotted against time.

■ Repolarization occurs as K^+ channels open and K^+ diffuses out of the cell. At the same time, Ca^{2+} channels close. These events restore the membrane to its original polarization, except that the positions of K^+ and Na^+ on each side of the sarcolemma are reversed.

■ A refractory period follows during which concentration of K^+ and Na^+ are actively restored to their appropriate side of the sarcolemma by Na^+/K^+ pumps. The muscle cell cannot contract again until Na^+ and K^+ are restored to their resting potential state. The refractory period of cardiac muscle is dramatically longer than that of skeletal muscle. This prevents tetanus from occurring and ensures that each contraction is followed by enough time to allow the heart chamber to refill with blood before the next contraction.

Electrocardiogram

Electrical currents generated by the heart during the cardiac cycle can be detected on the surface of the body by the electrodes of an electrocardiograph. A recording of these currents, called an electrocardiogram (ECG or EKG), represents a sum of all the concurrent action potentials produced by the heart as detected by the 12 electrodes of the electrocardiograph. A single cardiac cycle produces a distinctive wave pattern, where peaks and valleys are indicated by the letters P, Q, R, S, and T (see Figure 14-2). An interpretation of the major characteristics of the ECG follows:

■ The P wave is a small wave that represents the depolarization of the atria. During this wave, the muscles of the atria are contracting.

■ The QRS complex is a rapid down-up-down movement. The upward movement produces a tall peak, indicated by R. The QRS complex represents the depolarization of the ventricles.

■ The T wave represents the repolarization of the ventricles. Electrical activity generated by the repolarization of the atria is concealed by the QRS complex.

Figure 14-2 Different periods of the cardiac cycle: (a) electrocardiogram, (b) heart sounds, (c) valves, (d) pressure, and (e) volume of left ventricle.

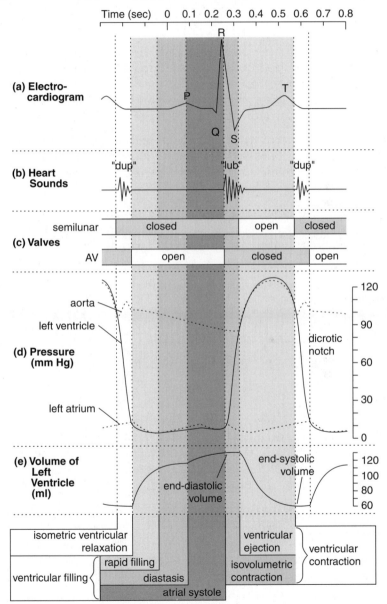

The Cardiac Cycle

The **cardiac cycle** describes all the activities of the heart through one complete heartbeat—that is, through one contraction and relaxation of both the atria and ventricles. A contraction event (of either the atria or ventricles) is referred to as **systole,** and a relaxation event is referred to as a **diastole.** The cardiac cycle includes a description of the systolic and diastolic activities of the atria and ventricles, the blood volume and pressure changes within the heart, and the action of the heart valves. A description of each period of the cardiac cycle follows (also refer to Figure 14-2):

- The isovolumetric ventricular relaxation is the period during which the ventricles are relaxed and both AV and semilunar valves are still closed. The volume of the ventricles remains unchanged (isovolumetric) during this period.

- Ventricular filling begins as the AV valves open and blood fills the ventricles. The ventricles remain in diastole during this period. The filling of the ventricles can be described as three successive events:

   Rapid ventricular filling occurs as blood flows into the empty and relaxed ventricles. Volume of the ventricles increases rapidly.

   Diastasis is a slower filling event than that of the preceding because most of the volume of the ventricle is already occupied by blood.

   Atrial systole (and the P wave of the ECG) occurs and forces the remaining blood from the atria into the ventricles. The blood volume at the end of this interval is called the end-diastolic volume (EDV).

- Ventricular contraction (ventricular systole) begins as the action potential from the AV node enters the ventricles, the ventricles depolarize, and the QRS complex is observed on the ECG. The following intervals during this phase are observed:

   Isovolumetric contraction occurs when the AV valves are forced shut. During this brief period, while the semilunar valves are still closed, the volume of the ventricles remains unchanged.

   Ventricular ejection occurs as the continuing contraction of the ventricles increases the pressure in the ventricles and forces the semilunar valves open. At this point, blood is forced out of the ventricles. This interval ends when the ventricles begin to relax, the blood in the aorta and pulmonary trunk begin to flow backward, and the semilunar valves, as a result, close. The closing of the semilunar valves causes a small increase in blood pressure visible as the dicrotic notch on a plot of blood pressure against time. The amount of blood remaining in the ventricles at this time is called the end-systolic volume (ESV).

The heart sounds associated with the beating of the heart can be heard by auscultating (listening to) the thorax with a stethoscope. The two major heart sounds, described as "lub-dup," originates from blood turbulence generated by the closing of the AV valves and the semilunar valves, respectively (refer to Figure 14-2). Abnormal heart sounds called murmurs are usually caused by improperly functioning valves.

Cardiac Output

The following variables are measures of the capacity of the heart:

- **Stroke volume** (SV) is the volume of blood ejected by each ventricle during a single contraction. The stroke volumes of the left and right ventricles are normally equal.

- **Heart rate** (HR) is the number of heartbeats per minute.

- **Cardiac output** (CO) is the volume of blood pumped out of the right or left ventricle per minute. CO = SV x HR.

Cardiac output varies widely with the health of the individual and the state of activity at the time of measurement. Cardiac output in exercising athletes may exceed their resting cardiac output seven times. The ratio between the maximum and resting cardiac output of an individual is the cardiac reserve. Note that cardiac output changes when either stroke volume or heart rate changes.

Stroke volume is regulated by the following three factors:

- Preload is the degree to which cardiac muscle cells are stretched by the blood entering the heart chambers. According to the Frank-Starling law of the heart, the more the chamber is stretched, the greater the force of its contraction. Because end-diastole volume (EDV) is a measure of how much blood enters the ventricles, the EDV is an indicator of ventricle preload.

- Contractility is the degree to which cardiac muscle cells contract as a result of extrinsic influences. Positive inotropic factors, such as certain hormones (epinephrine or thyroxin), drugs (digitalis), or elevated levels of Ca^{2+}, increase contractility, while negative inotropic factors, such as certain drugs (calcium channel blockers) or elevated levels of K^+, decrease contractility.

■ Afterload is a measure of the pressure that must be generated by the ventricles to force the semilunar valves open. The greater the afterload, the smaller the stroke volume. Arteriosclerosis (narrowing of the arteries) and high blood pressure increase afterload and reduce stroke volume.

Heart rate is regulated by the following three factors:

■ The autonomic nervous system may influence heart rate when the sympathetic nervous system stimulates cardiac muscle contractions or when the parasympathetic system inhibits cardiac muscle contractions.

■ Chemicals such as hormones and ions can influence heart rate. Epinephrine, secreted by the adrenal medulla, and thyroxin, secreted by the thyroid gland, increase heart rate. Abnormal blood concentrations of Na^+, K^+, and Ca^{2+} interfere with muscle contraction.

■ Other factors may also influence heart rate. These include age, gender, body temperature, and physical fitness.

Blood Vessels

The central opening of a blood vessel, the lumen, is surrounded by a wall consisting of three layers:

■ The tunica intima is the inner layer facing the blood. It is composed of an innermost layer of endothelium (simple squamous epithelium) surrounded by variable amounts of connective tissues.

■ The tunica media, the middle layer, is composed of smooth muscle with variable amounts of elastic fibers.

■ The tunica adventitia, the outer layer, is composed of connective tissue.

The cardiovascular system consists of three kinds of blood vessels that form a closed system of passageways:

■ Arteries carry blood away from the heart. The three kinds of arteries are categorized by size and function:

Elastic arteries (conducting arteries) are the largest arteries and include the aorta and other nearby branches. The tunica media of elastic arteries contains a large amount of elastic connective tissue, which enables the artery to expand as blood enters the lumen from the contracting heart. During relaxation of the heart, the elastic wall of the artery recoils to its original position, forcing blood forward and smoothing the jerky discharge of blood from the heart.

Muscular arteries (conducting arteries) branch from elastic arteries and distribute blood the various body regions. Abundant smooth muscle in the thick tunica media allows these arteries to regulate blood flow by vasoconstriction (narrowing of the lumen) or vasodilation (widening of the lumen). Most named arteries of the body are muscular arteries.

Arterioles are small, nearly microscopic, blood vessels that branch from muscular arteries. Most arterioles have all three tunics present in their walls, with considerable smooth muscle in the tunica media. The smallest arterioles consist of endothelium surrounded by a single layer of smooth muscle. Arterioles regulate the flow of blood into capillaries by vasoconstriction and vasodilation.

■ Capillaries are microscopic blood vessels with extremely thin walls. Only the tunica intima is present in these walls, and some walls consist exclusively of a single layer of endothelium. Capillaries penetrate most body tissues with dense interweaving networks called capillary beds. The thing walls of capillaries allow the diffusion of oxygen and nutrients out of the capillaries, while allowing carbon dioxide and wastes into the capillaries.

Metarterioles (precapillaries) are the blood vessels between arterioles and venules. Although metarterioles pass through capillary beds with capillaries, they are not true capillaries because metarterioles, like arterioles, have smooth muscle present in the tunica media. The smooth muscle of a metarteriole allows it to acts as a shunt to regulate blood flow into the true capillaries that branch from it. The thoroughfare channel, the tail end of the metarteriole that connects to the venule, lacks smooth muscle.

True capillaries form the bulk of the capillary bed. They branch away from a metarteriole at its arteriole end and return to merge with the metarteriole at its venule end (thoroughfare channel).

Some true capillaries connect directly from an arteriole to a metarteriole or venule. Although the walls of true capillaries lack muscle fibers, they posses a ring of smooth muscle called a precapillary sphincter where they emerge from the metarteriole. The precapillary sphincter regulates blood flow through the capillary. There are three types of true capillaries:

Continuous capillaries have continuous, unbroken walls consisting of cells that are connected by tight junctions. Most capillaries are of this type.

Fenestrated capillaries have continuous walls between endothelial cells, but the cells have numerous pores (fenestrations) that increase their permeability. These capillaries are found in the kidneys, lining the small intestine, and in other areas where a high transfer rate of substances into or out of the capillary is required.

Sinusoidal capillaries (sinusoids) have large gaps between endothelial cells that permit the passage of blood cells. These capillaries are found in the bone marrow, spleen, and liver.

■ Veins carry blood toward the heart. The three kinds of veins are listed here in the order that they merge to form increasingly larger blood vessels:

Postcapillary venules, the smallest veins, form when capillaries merge as they exit a capillary bed. Much like capillaries, they are very porous, but with scattered smooth muscle fibers in the tunica media.

Venules form when postcapillary venules join. Although the walls of larger venules contain all three layers, they are still porous enough to allow white blood cells to pass.

Veins have walls with all three layers, but the tunica intima and tunica media are much thinner than in similarly sized arteries. Few elastic or muscle fibers are present. The wall consists of primarily of a well-developed tunica adventitia. Many veins, especially those in the limbs, have valves, formed from folds of the tunica intima, that prevent the backflow of blood.

Many regions of the body receive blood supplies from two or more arteries. The points where these arteries merge are called arterial anastomoses. Arterial anastomoses allow tissues to receive blood even after one of the arteries supplying blood has been blocked.

Blood Pressure

Hydrostatic pressure created by the heart forces blood to move through the arteries. Systolic blood pressure, the pressure measured during contraction of the ventricles, averages about 120 mm Hg in arteries of the systemic circulation (for healthy, young adults). The diastolic blood pressure, measured during ventricle relaxation, is about 80 mm Hg in these arteries. As blood travels through the arterial system, resistance from the walls of the blood vessels reduces the pressure and velocity of the blood (see Figure 14-3). Blood pressure drops sharply in the arterioles and falls to between 40 and 20 mm Hg in the capillaries. Blood pressure descends further in the venules and approaches zero in the veins.

Figure 14-3 As blood travels through the arterial system, resistance from the walls of the blood vessels reduces the pressure and velocity of the blood.

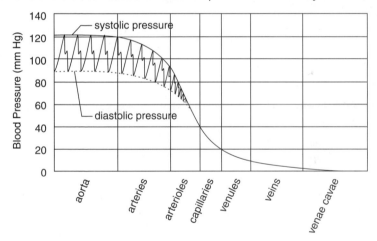

Because blood pressure is so low in venules and veins, two mechanisms assist the return of blood to the heart (venous return):

- The muscular pump arises from contractions of skeletal muscles surrounding the veins. The contractions squeeze the veins, forcing the blood to move forward, the only direction it can move when valves in the veins close to prevent backflow.

- The respiratory pump is created by the expansion and contraction of the lungs during breathing. During inspiration (inhaling), pressure in the abdominal region increases while pressure in the thoracic cavity decreases. These pressures act upon the veins passing through these regions. As a result, blood flows toward the heart as it moves from regions of higher pressure (the abdomen) to those of lower pressure (the chest and right atrium). When the pressures are reversed during expiration (exhaling), backflow in the veins is prevented by valves.

Control of Blood Pressure

Changes in blood pressure are routinely made in order to direct appropriate amounts of oxygen and nutrients to specific parts of body. For example, when exercise demands additional supplies of oxygen to skeletal muscles, blood delivery to these muscles increases, while blood delivery to the digestive organs decreases. Adjustments in blood pressure are also required when forces are applied to your body, such as when starting or stopping in an elevator.

Blood pressure can be adjusted by producing changes in the following variables:

■ Cardiac output can be altered by changing stroke volume or heart rate.

■ Resistance to blood flow in the blood vessel is most often altered by changing the diameter of the vessel (vasodilation or vasoconstriction). Changes in blood viscosity (its ability to flow) or in the length of the blood vessels (which increases with weight gain) can also alter resistance to blood flow.

The following mechanisms help regulate blood pressure:

■ The cardiovascular center provides a rapid, neural mechanism for the regulation of blood pressure by managing cardiac output or by adjusting blood vessel diameter. Located in the medullar oblongata of the brain stem, it consists of three distinct regions:

> The cardiac center stimulates cardiac output by increasing heart rate and contractility. These nerve impulses are transmitted over sympathetic cardiac nerves.

> The cardiac center inhibits cardiac output by decreasing heart rate. These nerve impulses are transmitted over parasympathetic vagus nerves.

The vasomotor center regulates blood vessel diameter. Nerve impulses transmitted over sympathetic motor neurons called vasomotor nerves enervate smooth muscles in arterioles throughout the body to maintain vasomotor tone, a steady state of vasoconstriction appropriate to the region.

■ The cardiovascular center receives information about the state of the body through the following sources.

Baroreceptors are sensory neurons that monitor arterial blood pressure. Major baroreceptors are located in the carotid sinus (an enlarged area of the carotid artery just above its separation from the aorta), the aortic arch, and the right atrium.

Chemoreceptors are sensory neurons that monitor levels of CO_2, O_2, and H^+ (pH). These neurons alert the cardiovascular center when levels of O_2 drop or levels of CO_2 and H^+ rise. Chemoreceptors are found in carotid bodies and aortic bodies located near the carotid sinus and aortic arch.

Higher brain regions, such as the cerebral cortex, hypothalamus, and limbic system, signal the cardiovascular center when conditions (stress, fight-or-flight response, hot or cold temperatures) require adjustments to the blood pressure.

■ The kidneys provide a hormonal mechanism for the regulation of blood pressure by managing blood volume.

The renin-angiotensis-aldosterone system of the kidneys regulates blood volume. In response to rising blood pressure, the juxtaglomerular cells in the kidneys secrete renin into the blood. Renin converts the plasma protein angiotensinogen to angiotensin I, which, in turn, is converted to angiotensisn II by the lungs. Angiotensin II activates two mechanisms that raise blood pressure.

Angiotensis II constricts blood vessels throughout the body (raising blood pressure by increasing resistance to blood flow). Constricted blood vessels reduce the amount of blood delivered to the kidneys, which decreases the kidneys' potential to excrete water (raising blood pressure by increasing blood volume).

Angiotensis II stimulates the adrenal cortex to secrete aldosterone, a hormone that reduces urine output by increasing retention of H_2O by the kidneys (increasing blood pressure by increasing blood volume).

■ Various substances influence blood pressure. Some important examples follow:

Epinephrine and norepinephrine, hormones secreted by the adrenal medulla, increase blood pressure by increasing heart rate and the contractility of the heart muscles and by causing vasoconstriction of arteries and veins. These hormones are secreted as part of the fight-or-flight response.

Antidiuretic hormone (ADH), a hormone secreted by the hypothalamus, increases blood pressure by stimulating the kidneys to retain H_2O (increasing blood pressure by increasing blood volume).

Arterial natriuretic (ANP), a hormone secreted by the atria of the heart, decreases the blood pressure by causing vasodilation and by stimulating the kidneys to excrete more water (decreasing blood pressure by reducing blood volume).

Nitric oxide (NO), secreted by endothelial cells, causes vasodilation.

Nicotine in tobacco increases blood pressure by stimulating sympathetic neurons to increase vasoconstriction and by stimulating the adrenal medulla to increase secretion of epinephrine and norepinephrine.

Alcohol decreases blood pressure by inhibiting the vasomotor center (causing vasodilation) and by inhibiting the release of ADH (increasing H_2O output, which decreases blood volume).

Blood Vessels of the Body

Figure 14-4 and 14-5 show the major arteries and veins of the body.

Figure 14-4 The major arteries in the body.

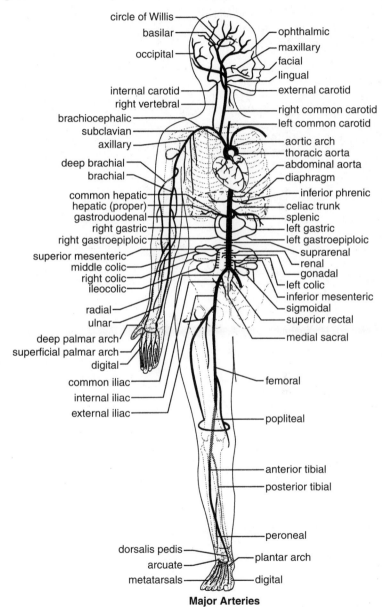

Major Arteries

Figure 14-5 The major veins in the body.

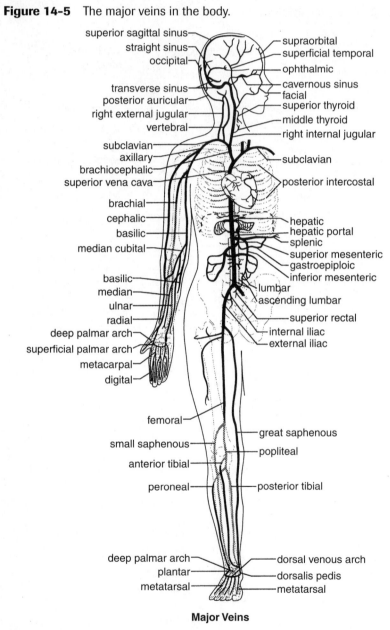

superior sagittal sinus
straight sinus
occipital

supraorbital
superficial temporal
ophthalmic

transverse sinus
posterior auricular
right external jugular
vertebral

cavernous sinus
facial
superior thyroid
middle thyroid
right internal jugular

subclavian
axillary
brachiocephalic
superior vena cava

subclavian

posterior intercostal

brachial
cephalic
basilic
median cubital

hepatic
hepatic portal
splenic
superior mesenteric
gastroepiploic
inferior mesenteric

basilic
median
ulnar
radial
deep palmar arch
superficial palmar arch
metacarpal
digital

lumbar
ascending lumbar
superior rectal
internal iliac
external iliac

femoral

small saphenous
anterior tibial

peroneal

great saphenous
popliteal

posterior tibial

deep palmar arch
plantar
metatarsal

dorsal venous arch
dorsalis pedis
metatarsal

Major Veins

Chapter Checkout

Q&A

1. Which variable is not a measure of the capacity of the heart?
   a. Cardiac Output (CO)
   b. Stroke Volume (SV)
   c. Heart Rate (HR)
   d. End-Systolic Volume (ESV)

2. True or False. The cardiac cycle describes all the activities of the heart through one complete heartbeat—that is, through one contraction and relaxation of both the atria and ventricles.

3. Which of the following is not a function of the cardiovascular system?
   a. Regulation of body temperature, fluid pH, and water content of cells.
   b. Absorbs lipids and fat-soluble materials from the digestive tract.
   c. Transport of nutrients, oxygen, and hormones to cells throughout the body.
   d. Protection of the body by white blood cells, antibodies, and complement proteins.

4. _____ are microscopic blood vessels with extremely thin walls.

5. True or False. Alcohol increases blood pressure by inhibiting the vasomotor center and by inhibiting the release of ADH.

Answers: 1. d 2. T 3. b 4. Capillaries 5. F

Chapter 15

THE LYMPHATIC SYSTEM

Chapter Check-In

❑ Knowing the three main functions of the lymphatic system

❑ Naming the lymphatic vessels and lymphoid cells

❑ Understanding the three categories of lymphatic tissues and organs

❑ Recognizing the functions and structure of the lymph nodes

❑ Listing the cells of the thymus and functions of the spleen

An important supplement to the cardiovascular system in helping to remove toxins from the body, the lymphatic system is also a crucial support of the immune system. Unlike blood, lymph only moves one way through your body, propelled by the action of nearby skeletal muscles. The lymph is pushed into the bloodstream for elimination. Appreciating the importance of the lymphatic system in filtering, recycling, and producing blood as well as filtering lymph, collecting excess fluids, and absorbing fat-soluble materials is important in the understanding of human physiology.

Lymphatic System Components

The **lymphatic system** consists of lymphatic vessels, a fluid called lymph, **lymph nodes,** the thymus, and the **spleen.** This system supplements and extends the cardiovascular system in the following ways:

■ The lymphatic system collects excess fluids and plasma proteins from surrounding tissues (interstitial fluids) and returns them to the blood circulation. Because lymphatic capillaries are more porous than blood capillaries, they are able to collect fluids, plasma proteins, and blood cells that have escaped from the blood. Within lymphatic vessels, this collected material forms a usually colorless fluid called lymph, which is transported to the neck, where it empties into the circulatory system.

■ The lymphatic system absorbs lipids and fat-soluble materials from the digestive tract.

■ The lymphatic system filters the lymph by destroying pathogens, inactivating toxins, and removing particulate matter. Lymph nodes, small bodies interspersed along lymphatic vessels, act as cleaning filters and as immune response centers that defend against infection.

The movement of lymph through lymphatic vessels is slow (3 liters/day) compared to blood flow (about 5 liters/minutes). Lymph does not circulate like blood, but moves in one direction from its collection in tissues to its return in the blood. There are no lymphatic pumps. Instead, lymph, much like blood in veins, is propelled forward by the action of the nearby skeletal muscles, the expansion and contraction of the lungs, and the contraction of the smooth muscle fibers in the walls of the lymphatic vessels. Valves in the lymphatic vessels prevent the backward movement of lymph.

Lymphatic Vessels

Lymphatic vessels occur throughout the body alongside arteries (in the viscera) or veins (in the subcutaneous tissue). They are absent from the central nervous system, bone marrow, teeth, and avascular tissues.

■ Lymph capillaries, the smallest lymphatic vessels, begin as dead-end vessels. They resemble blood capillaries, but are much more porous to surrounding fluids due to the following two features:

Valvelike openings form at the juncture of adjacent endothelial cells. Unlike the tightly joined endothelial cells that make up the walls of blood capillaries, those of lymph capillaries loosely overlap. When fluid pressure increases in surrounding regions, the overlapped cells separate, allowing fluids to enter the lymph capillary. When pressure inside the capillary exceeds the pressure outside, the spaces between the endothelial cells close, holding fluids inside the capillary.

Anchoring filaments attach the endothelial cells of the lymphatic vessels to surrounding collagen. When interstitial fluid pressure increases, the anchoring filaments prevent the endothelial cells from collapsing, keeping the spaces between the endothelial cells open.

■ Lacteals are specialized lymph capillaries that occur in the fingerlike projections (villi) that extend into the small intestine. Lacteals absorb lipids from the intestinal tract. The lymph within these capillaries, called chyle, has a creamy white color (rather than clear) due to the presence of fats.

- Lymphatic collecting vessels form as lymph capillaries merge. Collecting vessels have the following characteristics:

    Valves are present to prevent the backward flow of lymph (as in veins).

    The walls of collecting vessels consist of the same three tunics (layers) that characterize veins, but the layers are thinner and poorly defined.

- Lymphatic trunks form from the union of collecting vessels. The nine major trunks, draining lymph from regions for which they are named, are the lumbar, jugular, subclavian, and bronchomediastinal trunks, each of which occurs in pairs (left and right, for each side of the body), and a single intestinal trunk.

- Lymphatic ducts are the largest lymphatic vessels. These two ducts drain lymph into veins in the neck (the right and left subclavian veins at their junctures with the internal jugular veins). Valves in the lymphatic ducts at their junctures with the veins prevent the entrance of blood into the lymphatic vessels.

    The thoracic duct (left thoracic duct) collects lymph from the left side of the body and regions of the right side of body below the thorax. It begins at the cisterna chili, an enlarged region of the lymphatic vessel that forms following the union of the intestinal trunk and right and left lumbar trunks.

    The right thoracic duct collects lymph from the upper right side of the body (right arm and right regions of thorax, neck, and head), a much smaller area than that serviced by the thoracic duct.

Lymphoid Cells

Lymphatic (lymphoid) tissue is a kind of connective tissue. It consists of the following types of cells:

- **Lymphocytes** are white blood cells (leukocytes) that provide an immune response that attacks specific kinds of nonself cells and foreign substances (antigens). There are two major classes of lymphocytes:

T cells (T lymphocytes) originate in the bone marrow but mature in the thymus gland. T cells attack self cells that have been invaded by pathogens, abnormal self cells (such as cancerous cells), or nonself cells (such as those that might be introduced in an organ transplant).

B cells (B lymphocytes) originate and mature in the bone marrow. When B cells encounter an antigen (a toxin, virus, or bacterium), they produce plasma cells and memory cells. Plasma cells release antibodies that bind to the antigen and inactivate it. Memory cells circulate in the lymph and blood with the capacity to produce additional antigens for future encounters with the same antigen.

- **Macrophages** are enlarged monocytes (white blood cells) that engulf microbes and cellular debris.

- Reticular cells and their reticular fibers made from collagen and glycoproteins provide a network within which the lymphocytes and other cells reside.

Lymphatic Tissues and Organs

Lymphatic cells are organized into tissues and organs based upon how tightly the lymphatic cells are arranged and whether the tissue is encapsulated by a layer of connective tissue. Three general categories exist:

- *Diffuse, unencapsulated bundles of lymphatic cells.* This kind of lymphatic tissue consists of lymphocytes and macrophages associated with a reticular fiber network. It occurs in the lamina propria (middle layer) of the mucus membranes (mucosae) that line the respiratory and gastrointestinal tracts.

- *Discrete, unencapsulated bundles of lymphatic cells, called lymphatic nodules (follicles).* These bundles have clear boundaries that separate them from neighboring cells. Nodules occur within the lamina propria of the mucus membranes that line the gastrointestinal, respiratory, reproductive, and urinary tracts. They are referred to as mucosa-associated lymphoid tissue (MALT). The nodules contain lymphocytes and macrophages that protect against bacteria and other pathogens that may enter these passages with food, air, or urine. Nodules occur as solitary nodules, or they cluster as patches or aggregates. Here are the major clusters of nodules:

Peyer's patches are clusters of lymphatic nodules that occur in the mucosa that lines the ileum of the small intestine.

The tonsils are aggregates of lymphatic nodules that occur in the mucosa that lines the pharynx (throat). Each of the seven tonsils that form a ring around the pharynx are named for their specific region: A single pharyngeal tonsil (adenoid) in the rear wall of the nasopharynx, two palatine tonsils on each side wall of the oral cavity at its entrance in the throat, two lingual tonsils at the base of the tongue, and two small tubal tonsils in the pharynx at the entrance to the auditory tubes.

The appendix, a small fingerlike attachment to the beginning of the large intestine, is lined with aggregates of nodules.

- Encapsulated organs contain lymphatic nodules and diffuse lymphatic cells surrounded by a capsule of dense connective tissue. The three lymphatic organs are discussed in the following sections.

Lymph nodes

Lymph nodes are small, oval, or bean-shaped bodies that occur along lymphatic vessels. They are abundant where lymphatic vessels merge to form trunks, especially in the inguinal (groin), axillary (armpit), and mammary gland areas. Lymph flows into a node through afferent lymphatic vessels that enter the convex side of a node. It exits the node at the hilus, the indented region on the opposite, concave side of the node, through efferent lymphatic vessels. Efferent vessels contain valves that restrict lymph to movement in one direction out of the lymph node. The number of efferent vessels leaving the lymph node is fewer than the number of afferent vessels entering, slowing the flow of lymph through the node.

Lymph nodes perform three functions:

- They filter the lymph, preventing the spread of microorganisms and toxins that enter interstitial fluids.

- They destroy bacteria, toxins, and particulate matter through the phagocytic action of macrophages.

- They produce antibodies through the activity of B cells.

The structure of a lymph node is characterized by the following features:

- A capsule of dense connective tissue surrounds the lymph node.

- Trabeculae are projections of the capsule that extend into the node forming compartments. The trabeculae support reticular fibers that form a network that supports lymphocytes.

- The cortex is the dense, outer region of the node. It contains lymphatic nodules where B cells and macrophages proliferate.

- The medulla is the center of the node. Less dense than the surrounding cortex, the medulla primarily contains T cells.

- Medullary cords are strands of reticular fibers with lymphocytes and macrophages that extend from the cortex toward the hilus.

- Sinuses are passageways through the cortex and medulla through which lymph moves toward the hilus.

Thymus

The thymus is a bilobed organ located in the upper chest region between the lungs. It grows during childhood and reaches its maximum size of 40 g at puberty. It then slowly decreases in size as it is replaced by adipose and areolar connective tissue. By age 65, it weighs about 6 g.

Each lobe of the thymus is surrounded by a capsule of connective tissue. Lobules produced by trabeculae (inward extensions of the capsule) are characterized by an outer cortex and inner medulla. The following cells are present:

- Lymphocytes consist almost entirely of T cells.

- Epithelial-reticular cells resemble reticular cells, but do not form reticular fibers. Instead, these star-shaped cells form a reticular network by interlocking their slender cellular processes (extensions). These processes are held together by desmosomes, cell junctions formed by protein fibers.

- Epithelial-reticular cells produce thymosin and other hormones believed to promote the maturation of T cells.

- Thymic (Hassall's) corpuscles are dense, concentric layers of epithelial-reticular cells. Their function is unknown.

The function of the thymus is to promote the maturation of T lymphocytes. Immature T cells migrate through the blood from the red bone marrow to the thymus. Within the thymus, the immature T cells concentrate in the cortex where they continue their development. Mature T cells leave the thymus by way of blood vessels or efferent lymphatic vessels, migrating to other lymphatic tissues and organs where they become active (immunocompetent) in immune responses. The thymus does not provide a filtering function similar to lymph nodes (there are no afferent lymphatic vessels leading into the thymus), and unlike all other centers of lymphatic tissues, the thymus does not play a direct role in immune responses.

Blood vessels that permeate the thymus are surrounded by epithelial-reticular cells. These cells establish a protective blood-thymus barrier that prevents the entrance of antigens from the blood and into the thymus where T cells are maturing. Thus, an antigen-free environment is maintained for the development of T cells.

Spleen

Measuring about 12 cm (5 in) in length, the spleen is the largest lymphatic organ. It is located on the left side of the body between the diaphragm and stomach. Like other lymphatic organs, the spleen is surrounded by a capsule whose extensions into the spleen form trabeculae. The splenic artery, splenic vein, nerves, and efferent lymphatic vessels pass through the hilus of the spleen located on its slightly concave, upper surface. There are two distinct areas within the spleen:

■ White pulp consists of reticular fibers and lymphocytes in nodules that resemble the nodules of lymph nodes.

■ Red pulp consists of venous sinuses filled with blood. Splenic cords consisting of reticular connective tissue, macrophages, and lymphocytes form a mesh between the venous sinuses and act as a filter as blood passes between arterial vessels and the sinuses.

The functions of the spleen include the following:

■ *The spleen filters the blood.* Macrophages in the spleen remove bacteria and other pathogens, cellular debris, and aged blood cells. There are no afferent lymphatic vessels and, unlike lymph nodes, the spleen does not filter lymph.

■ *The spleen destroys old red blood cells and recycles their parts.* It removes the iron from heme groups and binds the iron to the storage protein.

■ *The spleen provides a reservoir of blood.* The diffuse nature of the red pulp retains large quantities of blood, which can be directed to the circulation when necessary. One third of the blood platelets are stored in the spleen.

■ *The spleen is active in immune responses.* T cells proliferate in the white pulp before returning to the blood to attack nonself cells when necessary. B cells proliferate in the white pulp, producing plasma cells and antibodies that return to the blood to inactivate antigens.

■ *The spleen produces blood cells.* Red and white blood cells are produced in the spleen during fetal development.

Chapter Checkout

Q&A

1. True or False. The spleen is the largest lymphatic organ.

2. Which of the following is not a function of the lymph nodes?

   a. They filter the lymph.
   b. They destroy old red blood cells.
   c. They destroy bacteria, toxins, and particulate matter.
   d. They produce antibodies.

3. The lymphatic system filters the lymph by destroying _____, inactivating toxins, and removing particulate matter.

4. Which of the following is a lymphatic vessel?

   a. Lacteals
   b. Lymphocytes
   c. Macrophages
   d. Reticular cells

5. True or False. The kind of lymphatic tissue that consists of lymphocytes and macrophages associated with a reticular fibernetwork is called lymphatic nodules.

Answers: 1. T 2. b 3. pathogens 4. a 5. F

Chapter 16

THE IMMUNE SYSTEM AND OTHER BODY DEFENSES

Chapter Check-In

❑ Listing nonspecific barriers and nonspecific defenses

❑ Recognizing the five main tasks of the immune system

❑ Understanding the major histocompatibility complex

❑ Describing the various kinds of lymphocytes

❑ Knowing the different immune responses and supplements to the immune response

The colossal task of keeping the body safe from outside and inside attacks of bacteria, virus, and other nasty critters belongs to the immune system. Your skin and mucous membranes are the first line of defense between invaders entering through the skin or through openings in the body. A second line of defense exists inside the body to challenge invaders that make it through the first line of defense. Sounds more like a war strategy, doesn't it? That's how your body treats anything that isn't supposed to be there—as an invader that must be destroyed.

Protecting Your Body

The internal environment of the human body provides attractive conditions for growth of bacteria, viruses, and other organisms. Although some of these organisms can live symbiotically within humans, many either cause destruction of cells or produce toxic chemicals. To protect against these foreign invaders, three lines of defense are employed: nonspecific barriers that deter the entrance of invaders and both nonspecific and specific defenses against invaders inside the body. A nonspecific defense is a rapid response to a wide range of pathogens. A specific defense, delivered by the

immune system, takes several days to mount and targets specific invaders that escape the attack of the nonspecific defense.

Nonspecific Barriers

The skin and mucous membranes provide a nonspecific first line of defense against invaders entering through the skin or through openings into the body. The first line of defense features the following mechanisms:

- Skin is a physical and hostile barrier covered with oily and acidic (pH from 3 to 5) secretions from sebaceous and sweat glands, respectively.

- Antimicrobial proteins (such as lysozyme, which breaks down the cell walls of bacteria) are contained in saliva, tears, and other secretions found on mucous membranes.

- Cilia that line the lungs serve to sweep invaders out of the lungs.

- Gastric juices of the stomach, by the action of hydrochloric acid or enzymes, kill most microbes.

- Symbiotic bacteria found in the digestive tract and the vagina out-compete many other organisms that could cause damage.

Nonspecific Defenses

The second line of defense consists of mechanisms or agents that indiscriminately challenge foreign invaders that are inside the body.

- **Phagocytes** are white blood cells (leukocytes) that engulf pathogens by phagocytosis. They include neutrophils, monocytes, and eosinophils. Monocytes enlarge into large phagocytic cells called macrophages.

- Natural killer cells (NK cells) are lymphocytes (white blood cells that mature in lymphoid tissues). NK cells kill pathogen-infected body cells or abdominal body cells (such as tumors).

- Complement is a group of about 20 proteins that "complement" defense reactions. These proteins help attract phagocytes to foreign cells and help destroy foreign cells by promoting cell lysis (breaking open the cell).

- Interferons (IFNs) are substances secreted by cells invaded by viruses and that stimulate neighboring cells to produce proteins that help them defend against the viruses. Certain IFNs (such as gamma-IFN) also amplify the activity of macrophages and natural killer cells.

■ The inflammatory response is a series of nonspecific events that occur in response to pathogens. The response typically produces redness, swelling, heat, and pain in the target area and often the area is disabled. When skin is damaged, for example, and bacteria, other organisms, or toxic substances enter the body, the following events occur:

> A chemical alarm is generated in the injured area. Injured cells and nearby circulating cells release chemicals that initiate defensive actions and sound an alarm to other defense mechanisms. These chemicals include histamine (mostly secreted by basophils, white blood cells found in connective tissue), kinins, prostaglandins (PGs), and complement.

> Vasodilation (dilation of blood vessels), stimulated by histamine and other chemicals, increases blood supply to the damaged area. This causes redness and an increase in local temperature. The increase in temperature stimulates white blood cells and makes the environment inhospitable to pathogens.

> Vascular permeability increases in response to alarm chemicals. As a result, white blood cells, clotting factors, and body fluids move more quickly through blood vessel walls and into the injured area. The increase in body fluids that results causes local edema (swelling). Edema may produce pain if nearby nerve endings experience pressure. Pain may also occur when nerve endings are exposed to bacterial toxins, kinins, and prostaglandins. (Aspirin reduces pain by inhibiting the production of prostaglandins.)

> Phagocytes arrive at the site of injury and engulf pathogens and damaged cells. Phagocytes find the site of injury by chemotaxis, the movement of cells in response to chemical gradients (provided here by alarm chemicals).

> Complement helps phagocytes engulf foreign cells, stimulates basophils to release histamine, and helps lyse foreign cells.

■ Fever is a total body response to infection characterized by elevated body temperature. An elevated temperature increases cellular metabolism (accelerating cellular repairs), amplifies the effect of alarm chemicals, and creates a hostile environment for bacteria. An excessively high fever may cause the breakdown of enzymes necessary for cellular metabolism, and death may result.

The Immune System

The immune system is the third line of defense. It consists of mechanisms and agents that target specific antigens (Ags). An **antigen** is any molecule, usually a protein or polysaccharide, that can be identified as foreign or nonself. It may be a toxin (injected into the blood by the sting of an insect, for example), a part of the protein coat of a virus, or a molecule unique to the plasma membranes of bacteria, protozoa, pollen, or other foreign cells. Once the antigen is recognized, an agent is released that targets the specific antigen. In the process of mounting a successful defense, the immune system accomplishes five tasks:

■ *Recognition.* The antigen or cell is recognized as nonself. To differentiate self from nonself, unique molecules on the plasma membrane of cells called the **major histocompatibility complex (MHC)** are used as a means of identification.

■ *Lymphocyte selection.* The primary defending cells of the immune system are certain white blood cells called lymphocytes. The immune system potentially possesses billions of lymphocytes, each equipped to target a different antigen. When an antigen, or nonself cell, binds to a lymphocyte, the lymphocyte proliferates, producing numerous daughter cells, all identical copies of the parent cell. This process is called clonal selection because the lymphocyte to which the antigen effectively binds is "selected" and subsequently reproduces to make clones, or identical copies, of itself.

■ *Lymphocyte activation.* The binding of an antigen or foreign cell to a lymphocyte may activate the lymphocyte and initiate proliferation. In most cases, however, a costimulator is required before proliferation begins. Costimulators may be chemicals or other cells.

■ *Destruction of the foreign substance.* Lymphocytes and antibodies destroy or immobilize the foreign substance. Nonspecific defense mechanisms (phagocytes, NK cells) help eliminate the invader.

■ *Memorization.* Long-lived "memory" lymphocytes are produced and can quickly recognize and respond to future exposures to the antigen or foreign cell.

Major Histocompatibility Complex

The major histocompatibility complex (MHC) (also called human leukocyte antigens, HLAs) is the mechanism by which the immune system is able to differentiate between self and nonself cells. The MHC is a collection of glycoproteins (proteins with a carbohydrate) that exist on the plasma membranes of nearly all body cells. The proteins of a single individual are unique, originating from 20 genes, with more than 50 variations per gene between individuals. Thus, it is extremely unlikely that two people, except for identical twins, will possess cells with the same set of MHC molecules.

The immune system is able to identify nonself cells by aberrations in the MHC displayed on the plasma membrane. There are two groups of MHC molecules, and each group generates different markings on the plasma membrane:

- MHC-I glycoproteins are produced by all body cells (except red blood cells). When a cell becomes cancerous or is invaded by a virus, unfamiliar proteins are synthesized in the cell. These proteins are endogenous antigens—that is, antigens produced inside the cell. Portions of these antigens are combined with MHC-I glycoproteins and, when displayed on the plasma membrane, indicate a nonself cell.

- MHC-II glycoproteins are produced only by antigen-presenting cells (APCs)—mostly macrophages and B cells. APCs actively ingest exogenous antigens, antigens that originate outside the cell. Exogenous antigens include viruses, toxins, pollen, or bacteria that are circulating in the blood, lymph, or body fluids. APCs break down the antigens and incorporate pieces of them with MHC-II proteins. This aberrant display of MHC markers is recognized as nonself.

Lymphocytes

The primary agents of the immune response are lymphocytes, white blood cells (leukocytes) that originate in the bone marrow (like all blood cells) but concentrate in lymphoid tissues such as the lymph nodes, the thymus gland, and the spleen. When lymphocytes mature, they become immunocompetent, or capable of binding with a specific antigen. An immunocompetent lymphocyte displays unique proteins on its plasma membrane that act as antigen receptors capable of binding to a specific antigen. Because all of the antigen receptors of an individual lymphocyte are identical, only a specific antigen can bind to an individual lymphocyte. The kind of antigen receptors displayed by a particular lymphocyte is determined by somatic recombination, a shuffling of gene segments during lymphocyte maturation. By mixing gene segments, more than one billion different antigen receptors can be generated.

Here are the various kinds of lymphocytes:

- B cells (B lymphocytes) are lymphocytes that originate and mature in the bone marrow. The antigen receptors of B cells bind to freely circulating antigens. When B cells encounter antigens that bind to their antigen binding sites, the B cells proliferate, producing two kinds of daughter cells, plasma cells and memory cells.

  Plasma cells are daughter cells of B cells. Each plasma cell releases antibodies, proteins that have the same antigen binding capability as the antigen receptors of its parent B cell. Antibodies circulate through the body binding to the specific antigens that stimulated the proliferation of plasma cells.

  Memory B cells are long-lived daughter cells of B cells that, like plasma cells, produce antibodies. However, memory cells do not release their antibodies in response to the immediate antigen invasion. Instead, the memory cells circulate in the body and respond quickly to eliminate any subsequent invasion by the same antigen. This mechanism provides immunity to many diseases after the first occurrence of the disease.

- T cells (T lymphocytes) are lymphocytes that originate in the bone marrow, but mature in the thymus gland. The antigen receptors of T cells bind to self cells that display foreign antigens (with MHC proteins) on their plasma membrane. When T cells bind to these aberrant self cells, they divide and produce the following kinds of daughter cells:

  Cytotoxic T cells (killer T cells) are activated when they recognize antigens that are mixed with the MHC-I proteins of self cells. Following activation, cytotoxic cells proliferate and destroy the recognized cells by producing toxins that puncture them, thus causing them to lyse.

  Helper T cells are activated when they recognize antigens that are mixed with the MHC-II proteins of self cells. Proliferation produces helper T cells that intensify antibody production of B cells. Helper T cells also secrete hormones called cytokines that stimulate the proliferation of B cells and T cells.

  Suppressor T cells are believed to be involved in winding down a successful immune response and in preventing the attachment of uninfected self cells.

  Memory T cells are long-lived cells possessing the same antigen receptors as their parent T cell. Like memory B cells, they provide a rapid defense to any subsequent invasion by the same antigen.

Antibodies

Antibodies are proteins that bind to specific antigens. B cells, located in lymphoid tissue, release the antibodies, which then circulate in the blood plasma, lymph, or extracellular fluids. Some antibodies migrate to other areas of the body, such as the respiratory tract or the placenta, or enter various body secretions, such as saliva, sweat, and milk. Additional properties of antibodies include the following:

■ There are five classes of antibodies (or immunoglobulins): IgA, IgD, IgE, IgG, and IgM. Antibodies circulating in the blood are primarily IgG, IgA, and IgM, IgD and a second form of IgM antibodies are found on the plasma membranes of B cells where they act as antigen receptors. IgE antibodies attach to basophils and mast cells (both white blood cells found in connective tissue) and induce them to secrete histamine.

■ The basic structure of an antibody is Y-shaped protein that consists of constant and variable regions. The variable regions are sequences of amino acids that differ among antibodies and give them specificity to antigens.

■ Antibodies inactivate antigens by binding to them and forming an antigen-antibody complex. Inactivation is followed by macrophage phagocytosis or lysis brought about by complement proteins. Inactivation may also cause agglutination (clumping) of antigens or foreign cells.

Costimulation

In some immune responses, a B cell or T cell becomes activated when an antigen or nonself cell binds to it. Activation then initiates proliferation. In most immune responses, however, activation requires the presence of a costimulator. That two signals, an antigen and a costimulator, are required to initiate the immune response ensures that healthy self cells are not destroyed. Costimulation may occur in two ways:

■ Cytokines, released by helper T cells and APCs, act as costimulators. Cytokines are protein hormones that influence cell growth. When a helper T cell becomes activated or an APC engulfs an antigen, the helper T cell or APC secretes a cytokine called interleukin.

■ Helper T cells and APCs act as costimulators. Activated T cells or APCs that display antigens activate B cells or T cells when they temporarily bind to them.

Humoral and Cell-Mediated Immune Responses

The immune system distinguishes two groups of foreign substances. One group consists of antigens that are freely circulating in the body. These include molecules, viruses, and foreign cells. A second group consists of self cells that display aberrant MHC proteins. Aberrant MHC proteins can originate from antigens that have been engulfed and broken down (exogenous antigens) or from virus-infected and tumor cells that are actively synthesizing foreign proteins (endogenous antigens). Depending upon the kind of foreign invasion, two different immune responses occur:

■ The humoral response (or antibody-mediated response) involves B cells that recognize antigens or pathogens that are circulating in the lymph or blood ("humor" is a medieval term for body fluid). The response follows this chain of events:

1. Antigens bind to B cells.

2. Interleukins or helper T cells costimulate B cells. In most cases, both an antigen and a costimulator are required to activate a B cell and initiate B cell proliferation.

3. B cells proliferate and produce plasma cells. The plasma cells bear antibodies with the identical antigen specificity as the antigen receptors of the activated B cells. The antibodies are released and circulate through the body, binding to antigens.

4. B cells produce memory cells. Memory cells provide future immunity.

■ The cell-mediated response involves mostly T cells and responds to any cell that displays aberrant MHC markers, including cells invaded by pathogens, tumor cells, or transplanted cells. The following chain of events describes this immune response:

1. Self cells or APCs displaying foreign antigens bind to T cells.

2. Interleukins (secreted by APCs or helper T cells) costimulate activation of T cells.

3. If MHC-I and endogenous antigens are displayed on the plasma membrane, T cells proliferate, producing cytotoxic T cells. Cytotoxic T cells destroy cells displaying the antigens.

4. If MHC-II and exogenous antigens are displayed on the plasma membrane, T cells proliferate, producing helper T cells. Helper T cells release interleukins (and other cytokines) which stimulate B cells to produce antibodies that bind to the antigens and stimulate nonspecific agents (NK and macrophages) to destroy the antigens.

Supplements to the Immune Response

Three important agents are used in medicine to supplement the immune response:

■ Antibiotics are chemicals derived from bacteria or fungi that are harmful to other microorganisms.

■ Vaccines are substances that stimulate the production of memory cells. Inactivated viruses or fragments of viruses, bacteria, or other microorganisms are used as vaccines. Once memory cells are formed, the introduction of the live microorganism will stimulate a swift response by the immune system before any disease can become established.

■ Passive immunity is obtained by transferring antibodies from an individual who previously had a disease to a newly infected individual. Newborn infants are protected by passive immunity through the transfer of antibodies across the placenta and by antibodies in breast milk.

Chapter Checkout

Q&A

1. True or False. Three important agents are used in medicine to supplement the immune response. These are antibiotics, vaccines, and passive immunity.

2. The skin and _____ _____ provide a _____ first line of defense against invaders entering through the skin or through openings into the body.

3. Which of the following is not a second line of defense that challenges foreign invaders that are inside the body?

   a. Fever
   b. Complement
   c. Antimicrobial proteins
   d. Inflammatory response

4. True or False. The major histocompatibility complex (MHC) is the mechanism by which the immune system is able to differentiate between self and nonself cells.

5. Which of the following lymphocytes originate and mature in the bone marrow?

   a. T cells
   b. B cells

Answers: 1. T 2. mucous membranes, nonspecific 3. c 4. T 5. b

Chapter 17

THE RESPIRATORY SYSTEM

Chapter Check-In

❏ Describing the respiration processes

❏ Recognizing the main structures of the respiratory system

❏ Knowing Boyle's Law on the mechanics of breathing

❏ Understanding gas exchange and transport

❏ Listing the respiratory centers

Out with the old and in with the new—that's what the respiratory system does, delivering air to the lungs, bringing oxygen into the body, and expelling the carbon dioxide back out into the air. Understanding the structure and intricacies of the respiratory system is vital to human anatomy. The respiratory system is made up of more than just the lungs as it also includes your nose, throat, larynx, windpipe, bronchi, alveolar ducts, and respiratory membrane.

Function of the Respiratory System

The function of the **respiratory system** is to deliver air to the lungs. Oxygen in the air diffuses out of the lungs and into the blood, while carbon dioxide diffuses in the opposite direction, out of the blood and into the lungs. Respiration includes the following processes:

■ Pulmonary ventilation is the process of breathing—inspiration (inhaling air) and expiration (exhaling air).

■ External respiration is the process of gas exchange between the lungs and the blood. Oxygen diffuses into the blood, while CO_2 diffuses from the blood into the lungs.

■ Gas transport, carried out by the cardiovascular system, is the process of distributing the oxygen throughout the body and collecting CO_2 and returning it to the lungs.

■ Internal respiration is the process of gas exchange between the blood, the interstitial fluids (fluids surrounding the cells), and the cells. Inside the cell, cellular respiration generates energy (ATP), using O_2 and glucose and producing waste CO_2.

Structure of the Respiratory System

The respiratory system is represented by the following structures, shown in Figure 17-1:

■ The nose consists of the visible external nose and the internal nasal cavity. The nasal septum divides the nasal cavity into right and left sides. Air enters two openings, the external nares (nostrils; singular, naris), and passes into the vestibule and through passages called meatuses. The bony walls of the meatuses, called concha, are formed by facial bones (the inferior nasal concha and the ethmoid bone). From the meatuses, air then funnels into two (left and right) internal nares. Hair, mucus, blood capillaries, and cilia that line the nasal cavity filter, moisten, warm, and eliminate debris from the passing air.

■ The pharynx (throat) consists of the following three regions, listed in order through which incoming air passes:

> The nasopharynx receives the incoming air from the two internal nares. The two auditory (Eustachian) tubes that equalize air pressure in the middle ear also enter here. The pharyngeal tonsil (adenoid) lies at the back of the nasopharynx.

> The oropharyrnx receives air from the nasopharynx and food from the oral cavity. The palatine and lingual tonsils are located here.

> The laryngopharynx passes food to the esophagus and air to the larynx.

Figure 17-1 A view of the entire respiratory system and the upper respiratory tract.

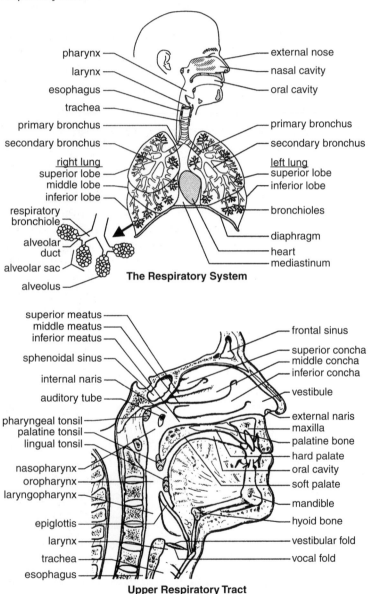

pharynx

larynx

esophagus

trachea

primary bronchus

secondary bronchus

right lung
superior lobe
middle lobe
inferior lobe

respiratory bronchiole

alveolar duct

alveolar sac

alveolus

external nose

nasal cavity

oral cavity

primary bronchus

secondary bronchus

left lung
superior lobe
inferior lobe

bronchioles

diaphragm

heart

mediastinum

The Respiratory System

superior meatus
middle meatus
inferior meatus

sphenoidal sinus

internal naris

auditory tube

pharyngeal tonsil
palatine tonsil
lingual tonsil

nasopharynx
oropharynx
laryngopharynx

epiglottis

larynx

trachea

esophagus

frontal sinus

superior concha
middle concha
inferior concha

vestibule

external naris
maxilla
palatine bone
hard palate
oral cavity
soft palate

mandible

hyoid bone

vestibular fold

vocal fold

Upper Respiratory Tract

■ The larynx receives air from the laryngopharynx. It consists of the following nine pieces of cartilage that are joined by membranes and ligaments, shown in Figure 17-2.

> The epiglottis, the first piece of cartilage of the larynx, is a flexible flap that covers the glottis, the upper region of the larynx, during swallowing to prevent the entrance of food.

> The thyroid cartilage protects the front of the larynx. A forward projection of this cartilage appears as the Adam's apple.

> The paired arytenoids cartilages in the rear are horizontally attached to the thyroid cartilage in the front by folds of mucous membranes. The upper vestibular folds (false vocal cords) contain muscle fibers that bring the folds together and allow the breath to be held during periods of muscular pressure on the thoracic cavity (straining while defecating or lifting a heavy object, for example). The lower vocal folds (true vocal cords) contain elastic ligaments that vibrate when skeletal muscles move them into the path of outgoing air. Various sounds, including speech, are produced in this manner.

> The cricoid cartilage, the paired cuneiform cartilages, and the paired corniculate cartilages are the remaining cartilages supporting the larynx.

■ The trachea (windpipe) is a flexible tube, 10 to 12 cm (4 inches) long and 2.5 cm (1 inch) in diameter, whose wall consists of four layers, as shown in Figure 17-2:

> The mucosa is the inner layer of the trachea. It contains mucus-producing goblet cells and pseudostratified ciliated epithelium. The movement of the cilia sweep debris away from the lungs toward the pharynx.

> The submucosa is a layer of areolar connective tissue that surrounds the mucosa.

> Hyaline cartilage forms 16 to 20 C-shaped rings that wrap around the submucosa. The rigid rings prevent the trachea from collapsing during inspiration.

> The adventitia is the outermost layer of the trachea. It consists of areolar connective tissue.

- The primary bronchi are two tubes that branch from the trachea to the left and right lungs.

- Inside the lungs, each primary bronchus divides repeatedly into branches of smaller diameters, forming secondary (lobar) bronchi, tertiary (segmental) bronchi, and numerous orders of bronchioles (1 mm or less in diameter), including terminal bronchioles (0.5 mm in diameter) and microscopic respiratory bronchioles. The wall of the primary bronchi are constructed like the trachea, but as the branches of the tree get smaller, the cartilaginous rings and the mucosa are replaced by smooth muscle.

Figure 17-2 Anterior and sagittal section of the larynx and the trachea.

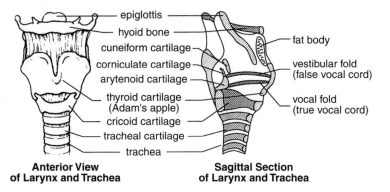

- Alveolar ducts are the final branches of the bronchial tree. Each alveolar duct has enlarged, bubblelike swellings along its length. Each swelling is called an alveolus, and a cluster of adjoining alveolar is called an alveolar sac. Some adjacent alveoli are connected by alveolar pores.

- The respiratory membrane consists of the alveolar and capillary walls. Gas exchange occurs across this membrane. Characteristics of this membrane follow:

  Type I cells are thin, squamous epithelial cells that constitute the primary cell type of the alveolar wall. Oxygen diffusion occurs across these cells.

  Type II cells are cuboidal epithelial cells that are interspersed among the type I cells. Type II cells secrete pulmonary surfactant (a phospholipid bound to a protein) that reduces the surface tension of the moisture that covers the alveolar walls. A reduction in surface tension permits oxygen to diffuse more easily into the moisture. A lower surface tension also prevents the moisture on opposite walls of an alveolus or alveolar duct from cohering and causing the minute airway to collapse.

Alveolar macrophage (dust cells) wander among the other cells of the alveolar wall removing debris and microorganisms.

A thin epithelial basement membrane forms the outer layer of the alveolar wall.

A dense network of capillaries surrounds each alveolus. The capillary walls consist of endothelial cells surrounded by a thin basement membrane. The basement membranes of the alveolus and the capillary are often so close that they fuse.

Lungs

The lungs are a pair of cone-shaped bodies that occupy the thorax (refer to Figure 17-1). The mediastinum, the cavity containing the heart, separates the two lungs. The left and right lungs are divided by fissures into two and three lobes, respectively. Each lobe of the lung is further divided into bronchopulmonary segments (each with a tertiary bronchus), which are further divided into lobules (each with a terminal bronchiole). Blood vessels, lymphatic vessels, and nerves penetrate each lobe.

Each lung has the following superficial features:

■ The apex and base identify the top and bottom of the lung, respectively.

■ The costal surface of each lung borders the ribs (front and back).

■ On the medial (mediastinal) surface, where each lung faces the other lung, the bronchi, blood vessels, and lymphatic vessels enter the lung at the hilus.

The pleura is a double membrane consisting of an inner pulmonary (visceral) pleura, which surrounds each lung, and an outer parietal pleura, which lines the thoracic cavity. The narrow space between the two membranes, the pleural cavity, is filled with pleural fluid, a lubricant secreted by the pleura.

Mechanics of Breathing

Boyle's law describes the relationship between the pressure (P) and the volume (V) of a gas. The law states that if the volume increases, then the pressure must decrease (or vice versa). This relationship is often written algebraically as PV= constant, or $P_1V_1 = P_2V_2$. Both equations state that the product of the pressure and volume remains the same. (The law applies only when the temperature does not change.)

Breathing occurs when the contraction or relaxation of muscles around the lungs changes the total volume of air within the air passages (bronchi, bronchioles) inside the lungs. When the volume of the lungs changes, the pressure of the air in the lungs changes in accordance with Boyle's law. If the pressure is greater in the lungs than outside the lungs, then air rushes out. If the opposite occurs, then air rushes in. Here is a summary of the process:

■ Inspiration occurs when the inspiratory muscles—that is, the diaphragm and the external intercostals muscles—contract. Contraction of the diaphragm (the skeletal muscle below the lungs) causes an increase in the size of the thoracic cavity, while contraction of the external intercostals muscles elevates the ribs and sternum. Thus, both muscles cause the lungs to expand, increasing the volume of their internal air passages. In response, the air pressure inside the lungs decreases below that of air outside the body. Because gases move from regions of high pressure to low pressure, air rushes into the lungs.

■ Expiration occurs when the diaphragm and external intercostal muscles relax. In response, the elastic fibers in lung tissue cause the lungs to recoil to their original volume. The pressure of the air inside the lungs then increases above the air pressure outside the body, and air rushes out. During high rates of ventilation, expiration is facilitated by contraction of the expiratory muscles (the intercostals muscles and the abdominal muscles).

Lung compliance is a measure of the ability of the lungs and thoracic cavity to expand. Due to the elasticity of lung tissue and the low surface tension of the moisture in the lungs (from the surfactant), the lungs normally have high compliance.

Lung Volumes and Capacities

The following terms describe the various lung (respiratory) volumes:

- The tidal volume (TV), about 500 ml, is the amount of air inspired during normal, relaxed breathing.

- The inspiratory reserve volume (IRV), about 3,100 ml, is the additional air that can be forcibly inhaled after the inspiration of a normal tidal volume.

- The expiratory reserve volume (ERV), about 1,200 ml, is the additional air that can be forcibly exhaled after the expiration of a normal tidal volume.

- Residual volume (RV), about 1,200 ml, is the volume of air still remaining in the lungs after the expiratory reserve volume is exhaled.

Summing specific lung volumes produces the following lung capacities:

- The total lung capacity (TLC), about 6,000 ml, is the maximum amount of air that can fill the lungs (TLC = TV + IRV + ERV + RV).

- The vital capacity (VC), about 4,800 ml, is the total amount or air that can be expired after fully inhaling (VC = TV + IRV + ERV = approximately 80% TLC).

- The inspiratory capacity (IC), about 3,600 ml, is the maximum amount of air that can be inspired (IC = TV + IRV).

- The functional residual capacity (FRC), about 2,400 ml, is the amount of air remaining in the lungs after a normal expiration (FRC = RV + ERV).

Some of the air in the lungs does not participate in gas exchange. Such air is located in the anatomical dead space within bronchi and bronchioles—that is, outside the alveoli.

Gas Exchange

In a mixture of different gases, each gas contributes to the total pressure of the mixture. The contribution of each gas, called the partial pressure, is equal to the pressure that the gas would have if it were alone in the enclosure. **Dalton's law** states that the sum of the partial pressures of each gas in a mixture is equal to the total pressure of the mixture.

The following factors determine the degree to which a gas will dissolve in a liquid:

■ *The partial pressure of the gas.* According to **Henry's law,** the greater the partial pressure of a gas, the greater the diffusion of the gas into the liquid.

■ *The solubility of the gas.* The ability of a gas to dissolve in a liquid varies with the kind of gas and the liquid.

■ *The temperature of the liquid.* Solubility decreases with increasing temperature.

Gas exchange occurs in the lungs between alveoli and blood plasma and throughout the body between plasma and interstitial fluids. The following factors facilitate diffusion of O_2 and CO_2 at these sites:

■ *Partial pressures and solubilities.* Poor solubility can be offset by a high partial pressure (or vice versa). Compare the following characteristics of O_2 and CO_2:

> Oxygen. The partial pressure of O_2 in the lungs is high (air is 21% O_2), but is solubility poor.

> Carbon dioxide. The partial pressure of CO_2 in air is extremely low (air is only 0.04% CO_2), but its solubility in plasma is about 24 times that of O_2.

■ *Partial pressure gradients.* A gradient is a change in some quantity from one region to another. Diffusion of a gas into a liquid (or the reverse) occurs down a partial pressure gradient—that is, from a region of higher partial pressure to a region of lower partial pressure. For example, the strong partial pressure gradient for O_2 (pO_2) from alveoli to deoxygenated blood (105 mm Hg in alveoli versus 40 mm Hg in blood) facilitates rapid diffusion.

■ *Surface area for gas exchange.* The expansive surface area of the lungs promotes extensive diffusion.

■ *Diffusion distance.* Thin alveolar and capillary walls increase the rate of diffusion.

Gas Transport

Oxygen is transported in the blood in two ways:

■ A small amount of O_2 (1.5 percent) is carried in the plasma as a dissolved gas.

- Most oxygen (98.5 percent) carried in the blood is bound to the protein hemoglobin in red blood cells. A fully saturated oxyhemoglobin (HbO_2) has four O_2 molecules attached. Without oxygen, the molecule is referred to as deoxygemoglobin (Hb).

The ability of hemoglobin to bind to O_2 is influenced by the partial pressure of oxygen. The greater the partial pressure of oxygen in the blood, the more readily oxygen binds to Hb. The oxygen-hemoglobin dissociation curve, shown in Figure 17-3, shows that as pO_2 increases toward 100 mm Hg, Hb saturation approaches 100%. The following four factors decrease the affinity, or strength of attraction, of Hb for O_2 and result in a shift of the O_2-Hb dissociation curve to the right:

- Increase in temperature.

- Increase in partial pressure of CO_2 (pCO_2).

- Increase in acidity (decrease in pH). The decrease in affinity of Hb for O_2, called the Bohr effect, results when H^+ binds to Hb.

- Increase in BPG in red blood cells. BPG (bisphosphoglycerate) is generated in red blood cells when they produce energy from glucose.

Figure 17-3 The oxygen-hemoglobin dissociation curve.

Carbon dioxide is transported in the blood in the following ways.

■ A small amount of CO_2 (8 percent) is carried in the plasma as a dissolved gas.

■ Some CO_2 (25 percent) binds to Hb in red blood cells forming carbaminohemoglobin ($HbCO_2$). (The CO_2 binds to a place different from that of O_2.)

■ Most CO_2 (65 percent) is transported as dissolved bicarbonate ions (HCO_3^- in the plasma. The formation of HCO3-, however, occurs in the red blood cells, where the formation of carbonic acid (H2CO3-) is catalyzed by the enzyme carbonic anhydrase, as follows.

$$CO_2 + H_2O \text{ fl‡ } H_2CO_3 \text{ fl‡ } H^+ + HCO_3.$$

Following their formation in the red blood cells, most H^+ bind to hemoglobin molecules (causing the Bohr effect) while the remaining H^+ diffuse back into the plasma, slightly decreasing the pH of the plasma. The HCO_3. ions diffuse back into the plasma as well. To balance the overall increase in negative charges entering the plasma, chloride ions diffuse in the opposite direction, from the plasma to the red blood cells (chloride shift).

Control of Respiration

Respiration is controlled by these areas of the brain that stimulate the contraction of the diaphragm and the intercostal muscles. These areas, collectively called respiratory centers, are summarized here:

■ The medullary inspiratory center, located in the medullar oblongata, generates rhythmic nerve impulses that stimulate contraction of the inspiratory muscles (diaphragm and external intercostal muscles). Normally, expiration occurs when these muscles relax, but when breathing is rapid, the inspiratory center facilitates expiration by stimulating the expiratory muscles (internal intercostal muscles and abdominal muscles).

■ The pheumotaxic area, located in the pons, inhibits the inspiratory center, limiting the contraction of the inspiratory muscles, and preventing the lungs from overinflating.

■ The apneustic area, also located in the pons, stimulates the inspiratory center, prolonging the contraction of inspiratory muscles.

The respiratory centers are influenced by stimuli received from the following three groups of sensory neurons:

- Central chemoreceptors (nerves of the central nervous system), located in the medulla oblongata, monitor the chemistry of cerebrospinal fluid. When CO_2 from the plasma enters the cerebrospinal fluid, it forms $HCO_3.$ and H+, and the pH of the fluid drops (becomes more acidic). In response to the decrease in pH, the central chemoreceptors stimulate the respiratory center to increase the inspiratory rate.

- Peripheral chemoreceptors (nerves of the peripheral nervous system), located in aortic bodies in the wall of the aortic arch and in carotid bodies in the walls of the carotid arteries, monitor the chemistry of the blood. An increase in pH or pCO_2, or decrease in pO_2, causes these receptors to stimulate the respiratory center.

- Stretch receptors in the walls of bronchi and bronchioles are activated when the lungs expand to their physical limit. These receptors signal the respiratory center to discontinue stimulation of the inspiratory muscles, allowing expiration to begin. This response is called the inflation (Hering-Breur) reflex.

Chapter Checkout

Q&A

1. Which of the following is not a process of respiration?
    a. Secretion
    b. Pulmonary ventilation
    c. External respiration
    d. Gas transport

2. True or False. Boyle's Law states that if the volume increases, then the pressure must increase.

3. Most oxygen carried in the blood is bound to the protein _____.

4. Which of the following is not a respiratory center of the brain?
    a. Pneumonic area
    b. Stretch receptors
    c. Medullary inspiratory center
    d. Apneustic area

5. The _____ _____ are two tubes that branch from the trachea to the left and right lungs.

Answers: 1. a 2. F 3. hemoglobin 4. c 5. primary bronchi

Chapter 18

THE DIGESTIVE SYSTEM

Chapter Check-In

❑ Describing the processes involved in the treatment of food

❑ Understanding the structure of the GI tract wall

❑ Summarizing the groups of molecules encountered during digestion

❑ Knowing the basic parts of the digestive system and their main functions

❑ Listing important hormones and their effects on the regulation of digestion

If you've ever wondered what happens to food after you eat it, this is the chapter where you find out that information. Your digestive system is made up of two major parts and it has two main functions: digestion and absorption. This complex system of breaking down food and absorbing the vital nutrients is something you should have a good understanding of in order to understand human physiology.

Function of the Digestive System

The function of the **digestive system** is digestion and absorption. Digestion is the breakdown of food into small molecules, which are then absorbed into the body. The digestive system is divided into two major parts:

■ The **gastrointestinal (GI) tract** (alimentary canal) is a continuous tube with two openings, the mouth and the anus. It includes the mouth, pharynx, esophagus, stomach, small intestine, and large intestine. Food passing through the internal cavity, or lumen, of the GI tract does not technically enter the body until it is absorbed through the walls of the GI tract and passes into blood or lymphatic vessels.

- Accessory organs include the teeth and tongue, salivary glands, liver, gallbladder, and pancreas.

The treatment of food in the digestive system involves the following seven processes:

- Ingestion is the process of eating.

- Propulsion is the movement of food along the digestive tract. The major means of propulsion is **peristalsis,** a series of alternating contractions and relaxations of smooth muscle that lines the walls of the digestive organs and that forces food to move forward.

- Secretion of digestive enzymes and other substances liquefies, adjusts the pH of, and chemically breaks down the food.

- Mechanical digestion is the process of physically breaking down food into smaller pieces. This process begins with the chewing of food and continues with the muscular churning of the stomach. Additional churning occurs in the small intestine through muscular constriction of the intestinal wall. This process, called segmentation, is similar to peristalsis, except that the rhythmic timing of the muscle constrictions forces the food backward and forward rather than forward only.

- Chemical digestion is the process of chemically breaking down food into simpler molecules. The process is carried out by enzymes in the stomach and small intestines.

- Absorption is the movement of molecules (by passive diffusion or active transport) from the digestive tract to adjacent blood and lymphatic vessels. Absorption is the entrance of the digested food into the body.

- Defecation is the process of eliminating undigested material through the anus.

Structure of the GI Tract Wall

The digestive tract, from the esophagus to the anus, is characterized by a wall with four layers, or tunics. Here are the layers, from the inside of the tract to the outside:

- The mucosa is a mucous membrane that lines the inside of the digestive tract from mouth to anus. Depending upon the section of the digestive tract, it protects the GI tract wall, secretes substances, and absorbs the end products of digestion. It is composed of three layers:

The epithelium is the innermost layer of the mucosa. It is composed of simple columnar epithelium or stratified squamous epithelium. Also present are goblet cells that secrete mucus that protects the epithelium from digestion and endocrine cells that secrete hormones into the blood.

The lamina propria lies outside the epithelium. It is composed of areolar connective tissue. Blood vessels and lymphatic vessels present in this layer provide nutrients to the epithelial layer, distribute hormones produced in the epithelium, and absorb end products of digestion from the lumen. The lamina propria also contains the mucosa-associated lymphoid tissue (MALT), nodules of lymphatic tissue bearing lymphocytes and macrophages that protect the GI tract wall from bacteria and other pathogens that may be mixed with food.

The muscularis mucosae, the outer layer of the mucosa, is a thin layer of smooth muscle responsible for generating local movements. In the stomach and small intestine, the smooth muscle generates folds that increase the absorptive surface area of the mucosa.

■ The submucosa lies outside the mucosa. It consists of areolar connective tissue containing blood vessels, lymphatic vessels, and nerve fibers.

■ The muscularis (muscularis externa) is a layer of muscle. In the mouth and pharynx, it consists of skeletal muscle that aids in swallowing. In the rest of the GI tract, it consists of smooth muscle (three layers in the stomach, two layers in the small and large intestines) and associated nerve fibers. The smooth muscle is responsible for movement of food by peristalsis and mechanical digestion by segmentation. In some regions, the circular layer of smooth muscle enlarges to form sphincters, circular muscles that control the opening and closing of the lumen (such as between the stomach and small intestine).

■ The serosa is a serous membrane that lines the outside of an organ. The following serosae are associated with the digestive tract:

The adventitia is the serous membrane that lines the esophagus.

The visceral peritoneum is the serous membrane that lines the stomach, large intestine, and small intestine.

The mesentery is an extension of the visceral peritoneum that attaches the small intestine to the rear abdominal wall.

The mesocolon is an extension of the visceral peritoneum that attaches the large intestine to the rear of the abdominal wall.

The parietal peritoneum lines the abdominopelvic cavity (abdominal and pelvic cavities). The abdominal cavity contains the stomach, small intestine, large intestine, liver, spleen, and pancreas. The pelvic cavity contains the urinary bladder, rectum, and internal reproductive organs.

Digestive Enzymes

During digestion, four different groups of molecules are commonly encountered. Each is broken down into its molecular components by specific enzymes:

- Complex carbohydrates, or polysaccharides, (such as starches) are broken down into oligosaccharides (consisting of two to ten linked monosaccharides), disaccharaides (such as maltose), or individual monosaccharides (such as glucose or fructose). Enzymes called amylases break down starch.

- Proteins are broken down into short chains of amino acids (peptides) or individual amino acids by enzymes called proteases.

- Fats (lipids) are broken down into glycerol and fatty acids by enzymes called lipases.

- Nucleic acids are broken down into nucleotides by enzymes called nucleases.

A summary of enzymes and their substrate (substances upon which enzymes operate) appear in Table 18-1.

Table 18-1   Enzymes and Their Substrates

Enzyme	Substrate	Products of Enzyme Activity
Saliva		
salivary amylase	starches	maltose, oligosaccharides
Gastric juice (chief cells of stomach)		
pepsin	proteins	peptides

(continued)

Table 18-1 *(continued)*

Enzyme	Substrate	Products of Enzyme Activity
Pancreatic juice (acinar cells of pancreas)		
pancreatic amylase	starches	maltose, oligosaccharides
trypsin	proteins	peptides
chymotrypsin	proteins	peptides
carboxypeptidase	proteins	peptides, amino acids
pancreatic lipase	fats	fatty acids, monoglycerides
nucleases	RNA & DNA	nucleotides
Brush border (absorptive cells of small intestine)		
dextrinase	oligosaccharides	glucose
maltase	maltose	glucose
sucrase	sucrose	glucose & fructose
lactase	lactose	glucose & galactose
aminopeptidase	peptides	peptides, amino acids
dipeptidase	dipeptides	amino acids
nucleosidases	nucleotides	nitrogen bases, ribose, deoxyribose & phosphates
phosphatases	nucleotides	nitrogen bases, ribose, deoxyribose & phosphates

The Mouth

The mouth (oral cavity, buccal cavity) is where food enters the digestive tract. The following features are found in the mouth:

- The vestibule is the narrow region between the cheeks and teeth and between the lips and teeth.

- The tongue defines the lower boundary of the mouth. It helps position the food during mastication (chewing) and gathers the chewed food into a ball, or bolus, in preparation for swallowing. The tongue is covered with papillae, small projections that help the tongue grip food. Many of the papillae bear taste buds.

- The palate defines the upper boundary of the mouth. The forward portion is the hard palate, hard because bone (maxillae and palatine) makes up this portion of it. Further back in the mouth, the soft palate consists of muscle and lacks any bone support. A conical muscular projection, the uvula, is suspended from the rear of the soft palate.

■ Saliva contains water (99.5 percent), digestive enzymes, lysozyme (an enzyme that kills bacteria), proteins, antibodies (IgA), and various ions. Saliva lubricates the mouth, moistens food during chewing, protects the mouth against pathogens, and begins the chemical digestion of food. Chemical digestion is carried out by the digestive enzyme salivary amylase, which breaks down polysaccharides (starch and glycogen) into short chains of glucose, especially the disaccharide maltose (which consists of two glucose molecules). Saliva is produced by the following glands:

> Three pairs of salivary glands, the parotid, submandibular, and sublingual, lie outside the mouth. They deliver their secretions to the mouth via ducts.

> Buccal glands are located in the mucosa that lines the mouth.

■ The teeth are embedded within sockets of the upper and lower jawbones (maxillae and mandible). Each tooth is surrounded by gum, or gingival, and held in its socket by a periodontal ligament. The 20 deciduous teeth (baby teeth or milk teeth) are eventually replaced by 32 permanent teeth. There are three types of teeth, based on shape and function:

> Incisors have a chisel-shaped edge suitable for biting off food.

> Canines (cuspids) are pointed fangs and are used for tearing food.

> Premolars (bicuspids) and molars have flat surfaces used for grinding and crushing.

A tooth has the following structural features:

> Dentin is a calcified tissue (like bone) that composes the bulk of the tooth.

> The crown is the portion of the tooth embedded in the bone.

> The neck is the region at the gum line where the crown and root meet.

> Enamel is the hard, nonliving material that covers the crown. Calcium compounds make the enamel the hardest substance in the body.

Cementum is the bonelike substance that covers the root and binds it to the periodontal ligament.

The pulp cavity is the central cavity inside the tooth. It contains blood vessels, nerves, and connective tissue (collectively called pulp).

The Pharynx

The pharynx, or throat, receives the food from the mouth during swallowing. From the mouth, it moves back and down into the oropharynx and then descends into the laryngopharynx. The food then passes into the esophagus.

The Esophagus

The esophagus is a 25-cm (10-inch) long tube that begins at the laryngopharynx and descends behind the trachea through the mediastinum (cavity between the lungs). It then passes through the diaphragm at an opening called the esophageal hiatus and connects to the stomach.

Food is forced through the esophagus toward the stomach by peristalsis. Two sphincter muscles, the upper esophageal sphincter at the top of the esophagus and the cardiac sphincter (lower esophageal sphincter) at the bottom of the esophagus, control the movement of food into and out of the esophagus.

Deglutition (Swallowing)

Swallowing, or deglutition, is divided into three phases:

- The buccal phase occurs voluntarily in the mouth when the tongue forces the food bolus back into the pharynx.

- The pharyngeal phase occurs involuntarily when food enters the pharynx, as follows:

   The soft palate and uvula fold upward and cover the nasopharynx to prevent the passage of food up and into the nasal cavity.

   The epiglottis, a flexible cartilaginous flap at the top of the larynx, folds down as the larynx rises. As a result, the opening to the larynx is covered, and food can pass only into the esophagus.

■ The esophageal phase occurs involuntarily in the esophagus. The esophageal sphincter, normally closed, opens to allow food to pass when the larynx rises during swallowing. When food reaches the lower end of the esophagus, the cardiac sphincter opens to allow the food to enter the stomach.

The Stomach

The stomach is J-shaped, baglike organ that expands to store food (Figure 18-1). Typical of that of the entire digestive tract, the wall of the stomach contains four layers. However, the inner layer, the mucosa, is modified for the specialized functions of the stomach. In particular, the innermost layer of the mucosa (facing the lumen) contains a layer of simple columnar epithelium consisting of goblet cells. Gastric pits on the surface penetrate deep into the layer, forming ducts whose walls are lined with various gastric glands. A summary of the glands in the mucosa follows:

■ Mucous surface cells are the goblet cells that make up the surface layer of the simple columnar epithelium. These cells secrete mucus, which protects the mucosa from the action of acid and digestive enzymes.

■ Mucous neck cells line the upper walls ("necks") of the ducts. Like the mucous surface cells, they secrete mucus.

■ Parietal (oxyntic) cells are scattered along the neck and lower walls of the ducts. They secrete hydrochloric acid (HC) and intrinsic factor. Intrinsic factor is necessary for the absorption of Vitamin B_{12} in the small intestine.

■ Chief (zymogenic) cells also line the lower walls of the ducts. They secrete pepsinogen, the inactive form of pepsin. Pepsin is a protease, an enzyme that breaks down proteins.

■ Enteroendocrine cells secrete various hormones that diffuse into nearby blood vessels. One important hormone, gastrin, stimulates other glands in the stomach to increase their output.

The first four glands listed in the preceding list are exocrine glands, whose secretions, collectively called gastric juice, enter the stomach and mix with food. The last gland is an endocrine gland, whose hormone secretions enter the blood supply.

Figure 18–1 The parts of the digestive system.

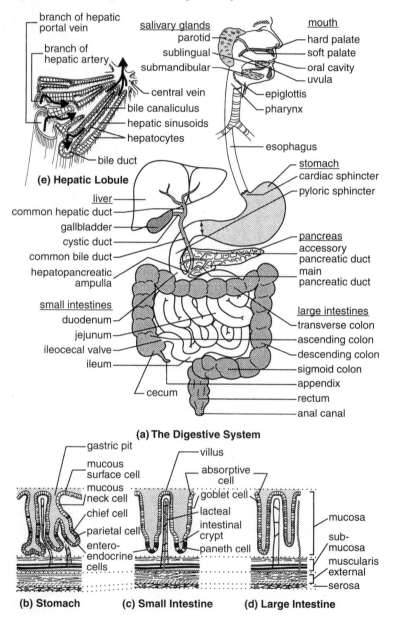

branch of hepatic portal vein

branch of hepatic artery

central vein
bile canaliculus
hepatic sinusoids
hepatocytes
bile duct

(e) Hepatic Lobule

salivary glands
parotid
sublingual
submandibular

mouth
hard palate
soft palate
oral cavity
uvula
epiglottis
pharynx

esophagus

stomach
cardiac sphincter
pyloric sphincter

liver
common hepatic duct
gallbladder
cystic duct
common bile duct
hepatopancreatic ampulla

pancreas
accessory pancreatic duct
main pancreatic duct

small intestines
duodenum
jejunum
ileocecal valve
ileum

large intestines
transverse colon
ascending colon
descending colon
sigmoid colon
appendix
rectum
anal canal

cecum

(a) The Digestive System

gastric pit
mucous surface cell
mucous neck cell
chief cell
parietal cell
entero-endocrine cells

villus
absorptive cell
goblet cell
lacteal
intestinal crypt
paneth cell

mucosa
sub-mucosa
muscularis
external
serosa

(b) Stomach   **(c) Small Intestine**   **(d) Large Intestine**

The stomach serves a variety of functions:

- *Storage.* Because of its accordionlike folds (called rugae), the wall of the stomach can expand to store two to four liters of material. Temporary storage is important because you eat considerably faster than you can digest food and absorb its nutrients.

- *Mixing.* The stomach mixes the food with water and gastric juice to produce a creamy medium called chyme.

- *Physical breakdown.* Three layers of smooth muscles (rather than the usual two) in the muscularis externa churn the contents of the stomach, physically breaking food down into smaller particles. In addition, HCl denatures (or unfolds) proteins and loosens the cementing substances between cells (of the food). The HCl also kills most bacteria that may accompany the food.

- *Chemical breakdown.* Proteins are chemically broken down by the enzyme pepsin. Chief cells, as well as other stomach cells, are protected from self-digestion because chief cells produce and secrete an inactive form of pepsin, pepsinogen. Pepsinogen is converted to pepsin by the HCl produced by the parietal cells. Only after pepsinogen is secreted into the stomach cavity can protein digestion begin. Once protein digestion begins, the stomach is protected by the layer of mucus secreted by the mucous cells.

- *Controlled release.* Movement of chyme into the small intestine is regulated by a valve at the end of the stomach, the pyloric sphincter.

The Small Intestine

The small intestine (small in diameter compared to the large intestine) is divided into three sections, as shown in Figure 18-1:

- The duodenum, about 25 cm (10 in) long, receives chyme from the stomach through the pyloric sphincter. Ducts that empty into the duodenum deliver pancreatic juice and bile from the pancreas and liver, respectively.

- The jejunum, about 2.5 m (8 ft) long, is the middle section of the small intestine.

- The ileum, about 3.6 m (12 ft) long, is the last section of the small intestine. It ends with the ileocecal valve (sphincter), which regulates the movement of chyme into the large intestine and prevents backward movement of material from the large intestine.

The functions of the small intestine include

- *Mechanical digestion.* Segmentation mixes the chyme with enzymes from the small intestine and pancreas. Bile from the liver separates fat into smaller fat globules. Peristalsis moves the chyme through the small intestine.

- *Chemical digestion.* Enzymes from the small intestine and pancreas break down all four groups of molecules found in food (polysaccharides, proteins, fats, and nucleic acids) into their component molecules.

- *Absorption.* The small intestine is the primary location in the GI tract for absorption of nutrients:

  Carbohydrates, proteins, nucleic acids, and water-soluble vitamins. The components of these molecules are absorbed by facilitated diffusion or active transport. They are then passed to blood capillaries.

  Vitamin B_{12}. Vitamin B_{12} combines with intrinsic factor (produced in the stomach) and is absorbed by receptor-mediated endocytosis. It is then passed to the blood capillaries.

  Lipids and fat-soluble vitamins. Because fat-soluble vitamins and the components of lipids are insoluble in water, they are packaged and delivered to cells within water-soluble clusters of bile salts called micelles. They are then absorbed by simple diffusion and, once inside the cells, mix with cholesterol and protein to form chylomicrons. The chylomicrons are then passed to the lymphatic capillaries. When the lymph eventually empties into the blood, the chylomicrons are broken down by lipoprotein lipase, and the breakdown products, fatty acids and glycerol, pass through blood capillary walls to be absorbed by various cells.

  Water and electrolytes. About 90 percent of the water in chyme is absorbed, as well as various electrolytes (ions), including Na^+, K^+, Cl^-, nitrates, calcium, and iron.

Modifications of the mucosa for its various specialized functions in the small intestine include the following:

- The plicae circulares (circular folds) are permanent ridges in the mucosa that encircle the inside of the small intestine. The ridges force the food to spiral forward. The spiral motion helps mix the chyme with the digestive juices.

■ Villi (singular, villus) are fingerlike projection that cover the surface of the mucosa, giving it a velvety appearance. They increase the surface area over which absorption and digestion occur. The spaces between adjacent villi lead to deep cavities at the bases of the villi called intestinal crypts (crypts of Lieberkühn). Glands that empty into the cavities are called intestinal glands, and the secretions are collectively called intestinal juice.

Microvilli are microscopic extensions of the outer surface of the absorptive cells that line each villus. Because of their brushlike appearance (microscopically), the microvilli facing the lumen form the brush border of the small intestine. Like the villi; the microvilli increase the surface area over which digestion and absorption take place.

The villi of the mucosa have the following characteristics:

■ An outer epithelial layer (facing the lumen) consists of the following cell types:

Absorptive cells, the primary cell type of the epithelial layer, synthesize digestive enzymes called brush border enzymes that become embedded in the plasma membranes around the microvilli. Various nutrients in the chyme that move over the microvilli are broken down by these brush border enzymes and subsequently absorbed. Refer to Table 18-1, earlier in this chapter, for a list of these enzymes.

Goblet cells, located throughout the epithelial layer, secrete mucus that helps protect the epithelial layer from digestion.

Enteroendocrine cells secrete hormones into blood vessels that penetrate the villus.

Paneth cells, located in the epithelial layer facing the intestinal crypts, secrete lysozyme, an enzyme that destroys bacteria.

■ An inner core of lamina propria (connective tissues) contains blood capillaries and a small lymphatic capillary called a lacteal.

The submucosa that underlies the mucosa of the small intestine bears the following modifications:

■ Brunner's (duodenal) glands, found only in the submucosa of the duodenum, secrete an alkaline mucus that neutralizes the gastric acid in the incoming chyme.

■ Peyer's patches (aggregated lymphatic nodules), found mostly in the submucosa of the ileum, are clusters of lymphatic nodules that provide a defensive barrier against bacteria.

Large Intestine

The large intestine is about 1.5 m (5 ft) long and is characterized by the following components:

■ The cecum is a dead-end pouch at the beginning of the large intestine, just below the ileocecal valve.

■ The appendix (vermiform appendix) is an 8 cm (3 in) long fingerlike attachment to the cecum that contains lymphoid tissue and serves immunity functions.

■ The colon, representing the greater part of the large intestine, consists of four sections: the ascending, transverse, descending, and sigmoid colons. At regular distances along the colon, the smooth muscle of the muscularis layer causes the intestinal wall to gather, producing a series of pouches called haustra. The epithelium facing the lumen of the colon is covered with openings of tubular intestinal glands that penetrate deep into the thick mucosa. The glands consist of absorptive cells that absorb water and goblet cells that secrete mucus. The mucus lubricates the walls of the large intestine to smooth the passage of feces.

■ The rectum is the last 20 cm (8 in) of the large intestine. The mucosa in the rectum forms longitudinal folds called anal columns.

■ The anal canal, the last 3 cm (1 in) of the rectum, opens to the exterior at the anus. An involuntary (smooth) muscle, the interior anal sphincter, and a voluntary (skeletal) muscle, the external anal sphincter, control the release of the feces through the anus.

The functions of the large intestine include

■ *Mechanical digestion.* Rhythmic contractions of the large intestine produce a form of segmentation called haustral contractions in which food residues are mixed and forced to move from one haustrum to the next. Peristaltic contractions produce mass movements of larger amounts of material.

- ■ *Chemical digestion.* Digestion occurs as a result of bacteria that colonize the large intestine. They break down indigestible material by fermentation, releasing various gases. Vitamin K and certain B vitamins are also produced by bacterial activity.

- ■ *Absorption.* Vitamins B and K, some electrolytes (Na^+ and Cl^-), and most of the remaining water is absorbed by the large intestine.

- ■ *Defecation.* Mass movement of feces into the rectum stimulates a defecation reflex that opens the internal anal sphincter. Unless the external and sphincter is voluntarily closed, feces are evacuated through the anus.

The Pancreas

The secretions of the pancreas, called pancreatic juice, include various enzymes, including pancreatic amylase (digestion of starch), trypsin, carboxypepiydase, and chymotrypsin (proteases), as well as pancreatic lipase (digestion of fats). Sodium bicarbonate is also produced, making the pancreatic juice alkaline. An alkaline solution neutralizes the HCl in the chyme and provides an optimal environment for the action of these enzymes.

Pancreatic juice is produced in clusters of exocrine cells called acini. The remaining cells in the pancreas (about 1 percent of the total) also form clusters (islets of Langerhans). These are the endocrine cells that produce the hormones insulin, glucagon, somatostatin, and pancreatic polypeptide.

Pancreatic juice collects in small ducts that merge to form two large ducts. The main pancreatic duct (duct of Wirsung) exits the pancreas and merges with the common bile duct from the liver and gallbladder. This combined duct, called the hepatopancreatic ampulla (ampulla of Vater), then enters the duodenum. A smaller, second duct that exits the pancreas, the accessory pancreatic duct (duct of Santorini), joins the duodenum directly.

The Liver and Gallbladder

The digestive function of the liver is to produce bile, which is then delivered to the duodenum to emulsify fats. Emulsification is the breaking up of fat globules into smaller fat droplets, increasing the surface area upon which fat-digesting enzymes (lipases) can operate. Because bile does not chemically change anything, it is not an enzyme. Bile is also alkaline, serving to help neutralize the HCl in the chyme.

Bile consists of bile salts, bile pigments, phospholipids (including lecithin), cholesterol, and various ions. The primary bile pigment, bilirubin, is an end product of the breakdown of hemoglobin from expended red blood cells. Although some of the bile is lost in the feces (bilirubin gives feces their brown color), much of the bile is reabsorbed by the small intestine and returned to the liver via the hepatic portal vein.

The liver performs numerous metabolic functions. Some of the most important follow.

- Bile is produced.

- Blood glucose is regulated. When blood glucose is high, the liver converts glucose to glycogen (glycogenesis) and stores the glycogen. When blood glucose is low, glycogen is broken down (glycogenolysis), and glucose is released into the blood.

- Proteins (including plasma proteins) and certain amino acids are synthesized.

- Ammonia (which is toxic) is converted to urea (less toxic) for elimination by the kidneys.

- Bacteria and expended red and white blood cells are broken down. From the red blood cells, Fe and globin are recycled, and bilirubin is secreted in the bile.

- Vitamins (A, D, and B_{12}) and minerals (including Fe from expended red blood cells) are stored.

- Toxic substances (drugs, poisons) and hormones are broken down.

The liver is composed of numerous functions units called lobules. Within each lobule, epithelial cells called hepatocytes are arranged in layers that radiate out from a central vein. Hepatic sinusoids are spaces that lie between groups of layers, while smaller channels called bile canaliculi separate other layers. Each of (usually) six corners of the lobule are occupied by three vessels: one bile duct and two blood vessels (a portal triad). The blood vessels are branches from the hepatic artery (carrying oxygenated blood) and from the hepatic portal vein (carrying deoxygenated but nutrient-rich blood from the small intestine).

Blood enters the liver through the hepatic artery and hepatic portal vein and is distributed to lobules. Blood flows into each lobule by passing through the hepatic sinusoid and collecting in the central vein. The central veins of all the lobules merge and exit the liver through the hepatic vein (not the hepatic portal vein).

Within the sinusoids, phagocytes called Kupffer cells (stellate reticuloendothelial cells) destroy bacteria and break down expended red and white blood cells and other debris. Hepatocytes that border the sinusoids also screen the incoming blood. They remove various substances from the blood, including oxygen, nutrients, toxins, and waste materials. From these substances, they produce bile, which they secrete into the bile canaliculi, which empty into bile ducts. Bile ducts from the various lobules merge and exit the liver as a single common hepatic duct. The common hepatic duct merges with the cystic duct from the gallbladder to form the common bile duct, which, in turn, merges with the pancreatic duct to form the hepatopancreatic ampulla. This last duct delivers the bile to the duodenum.

The gallbladder stores excess bile. When food is in the duodenum, bile flows readily from the liver and gallbladder into the duodenum. When the duodenum is empty, a sphincter muscle (sphincter of Oddi) closes the hepatopancreatic ampulla, and bile backs up and fills the gallbladder.

Regulation of Digestion

The activities of the digestive system are regulated by both hormones and neural reflexes. Four important hormones and their effects upon target cells follow.

- Gastrin is produced by enteroendocrine cells of the stomach mucosa. Effects include

    Stimulation of gastric juice (especially HCl) secretion by gastric glands.

    Stimulation of smooth muscle contraction in the stomach, small intestine, and large intestine, which increases gastric and intestinal motility.

    Relaxation of the pyloric sphincter, which promotes gastric emptying into the small intestine.

- Secretin is produced by the enteroendocrine cells of the duodenal mucosa. Effects include

    Stimulation of bicarbonate secretion by the pancreas, which neutralizes the acidity of chyme when released into the duodenum.

    Stimulation of bile production by the liver.

    Inhibition of gastric juice secretions and gastric motility, which, in turn, slows digestion in the stomach and retards gastric emptying.

- Cholecystokinin (CCK) is produced by enteroendocrine cells of the duodenal mucosa. Effects include

    Stimulation of bile release by the gallbladder.

    Stimulation of pancreatic juice secretion.

    Relaxation of the hepatopancreatic ampulla, which allows flow of bile and pancreatic juices into the duodenum.

- Gastric inhibitory peptide (GIP) is produced by enteroendocrine cells of the duodenal mucosa and causes the inhibition of gastric juice secretion and gastric motility, which, in turn, slows digestion in the stomach and retards gastric emptying.

The second regulatory agent of the digestive system is the nervous system. Stimuli that influence digestive activities may originate in the head, the stomach, or the small intestine. Based on these sites, there are three phases of digestive regulation:

- The cephalic phase comprises those stimuli that originate from the head: sight, smell, taste, or thoughts of food, as well as emotional states. In response, the following reflexes are initiated:

    Neural response. Stimuli that arouse digestion are relayed to the hypothalamus, which, in turn, initiates nerve impulses in the parasympathetic vagus nerve. These impulses innervate nerve networks of the GI tract (enteric nervous system), which promote contraction of smooth muscle (which causes peristalsis) and secretion of gastric juice. Stimuli that repress digestion (emotions of fear or anxiety, for example) innervate sympathetic fibers that suppress muscle contraction and secretion.

    General effects. The stomach prepares for the digestion of proteins.

- The gastric phase describes those stimuli that originate from the stomach. These stimuli include distention of the stomach (which activates stretch receptors), low acidity (high pH), and the presence of peptides. In response, the following reflexes are initiated:

    Neural response. Gastric juice secretion and smooth muscle contraction are promoted.

    Hormonal response. Gastrin production is promoted.

    General effects. The stomach and small intestine prepare for the digestion of chyme, and gastric emptying is promoted.

■ The intestinal phase describes stimuli originating in the small intestine. These include distention of the duodenum, high acidity (low pH), and the presence of chyme (especially fatty acids and carbohydrates). In response, the following reflexes are initiated:

Neural response. Gastric secretion and gastric motility are inhibited (enterogastric reflex). Intestinal secretions, smooth muscle contraction, and bile and pancreatic juice production are promoted.

Hormonal response. Production of secretin, CCK, and GIP is promoted.

General effects. Stomach emptying is retarded to allow adequate time for digestion (especially fats) in the small intestine. Intestinal digestion and motility are promoted.

Chapter Checkout

Q&A

1. Which of the following is the correct order of the processes involved in the treatment of food in the digestive tract?
   a. Ingestion, propulsion, defecation, chemical digestion, absorption, secretion, mechanical digestion
   b. Ingestion, propulsion, secretion, mechanical digestion, defecation, chemical digestion, absorption
   c. Ingestion, propulsion, secretion, mechanical digestion, chemical digestion, absorption, defecation
   d. Ingestion, secretion, propulsion, mechanical digestion, chemical digestion, absorption, defecation

2. True or False. The digestive tract, from the esophagus to the anus, is characterized by a wall with five layers, or tunics.

3. During digestion, four different groups of molecules are commonly encountered. These are complex carbohydrates, or polysaccharides, _____, fats, and _____ _____.

4. Which of the following is not part of the digestive system?
   a. Esophagus
   b. Stomach
   c. Large intestine
   d. Appendix

5. Four important hormones that help regulate digestion are gastrin, _____, cholecystokinin, and _____ _____ _____.

Answers: 1. c 2. F 3. proteins, nucleic acids 4. d 5. secretin, gastric inhibitory peptide (GIP)

Chapter 19

THE URINARY SYSTEM

Understanding how the urinary system helps maintain homeostasis by removing harmful substances from the blood and regulating water balance in the body is an important part of physiology. Your kidneys, which are the main part of the urinary system, are made up of millions of nephrons that act as individual filtering units and are complex structures themselves. The ureters, urethra, and urinary bladder complete this intricate system.

The **urinary system** helps maintain homeostasis by regulating water balance and by removing harmful substances from the blood. The blood is filtered by two **kidneys,** which produce urine, a fluid containing toxic substances and waste products. From each kidney, the urine flows through a tube, the ureter, to the urinary bladder, where it is stored until it is expelled from the body through another tube, the urethra.

Anatomy of the Kidneys

The kidneys are surrounded by three layers of tissue:

■ The renal fascia is a thin, outer layer of fibrous connective tissue that surrounds each kidney (and the attached adrenal gland) and fastens it to surrounding structures.

- The adipose capsule is a middle layer of adipose (fat) tissue that cushions the kidneys.

- The renal capsule is an inner fibrous membrane that prevents the entrance of infections.

Inside the kidney, three major regions are distinguished, as shown in Figure 19-1.

- The renal cortex borders the convex side.

- The renal medulla lies adjacent to the renal cortex. It consists of striated, cone-shaped regions called renal pyramids (medullary pyramids), whose peaks, called renal papillae, face inward. The unstriated regions between the renal pyramids are called renal columns.

- The renal sinus is a cavity that lies adjacent to he renal medulla. The other side of the renal sinus, bordering the concave surface of the kidney, opens to the outside through the renal hilus. The ureter, nerves, and blood and lymphatic vessels enter the kidney on the concave surface through the renal hilus. The renal sinus houses the renal pelvis, a funnel-shaped structure that merges with the ureter. The renal pelvis forms branches called major and minor calyces (singular, calyx) that extend into the renal medulla and abut the renal pyramids.

Blood and nerve supply

Because the major function of the kidneys is to filter the blood, a rich blood supply is delivered by the large renal arteries. The renal artery for each kidney enters the renal hilus and successively branches into segmental arteries, lobar arteries, and finally interlobar arteries that pass between the renal pyramids toward the renal cortex. The interlobar arteries then branch into the arcuate arteries, which curve as they pass along the junction of the renal medulla and cortex. Branches of the arcuate arteries, called interlobular arteries, penetrate the renal cortex, where they again branch into afferent arterioles, which enter the filtering mechanisms, or glomeruli, of the nephrons.

Blood leaving the nephrons exits the kidney through veins that trace the same path, in reverse, as the arteries that delivered the blood. Interlobular, arcuate, interlobar, and segmental veins successively merge and exit as a single renal vein.

Figure 19-1 (a) The urinary system, (b) The kidney, (c) cortical nephron and (d) juxtamedullary nephron of the kidneys.

(a)Urinary System

(b) Kidney

(c) Cortical Nephron **(d) Juxtamedullary Nephron**

Autonomic nerves from the renal plexus follow the renal artery into the kidney through the renal hilus. The nerve fibers follow the branching pattern of the renal artery and serve as vasomotor fibers that regulate blood volume. Sympathetic fibers constrict arterioles (decreasing urine output), while less numerous parasympathetic fibers dilate arterioles (increasing urine ouput).

The nephron

The kidney consists of over a million individual filtering units called **nephrons.** Each nephron consists of a filtering body, the renal corpuscle, and a urine-collecting and concentrating tube, the renal tubule. The renal corpuscle is an assemblage of two structures, the glomerulus and the Bowman's (glomerular) capsule, shown in Figure 19-1.

- The glomerulus is a dense ball of capillaries (glomerular capillaries) that branches from the afferent arteriole that enters the nephron. Because blood in the glomerular capillaries is under high pressure, substances in the blood that are small enough to pass through the pores (fenestrae, or endothelial fenestrations) in the capillary walls are forced out and into the encircling Bowman's capsule. The glomerular capillaries merge, and the remaining blood exits the glomerulus through the efferent arteriole.

- The Bowman's capsule is a cup-shaped body that encircles the glomerulus and collects the material (filtrate) that is forced out of the glomerular capillaries. The filtrate collects in the interior of the Bowman's capsule, the capsular (Bowman's) space, an area bounded by an inner visceral layer (that faces the glomerulus) and an outer parietal layer. The visceral layer consists of modified simple squamous epithelial cells called podocytes that project branches that bear fine processes called pedicels. The pedicels adjacent podocytes mesh to form a dense network that envelops the glomerular capillaries. Spaces between the pedicels, called filtration slits, are openings into the capsular space that allow filtrate to enter the Bowman's capsule.

- The renal tubule consists of three sections:

    The first section, the proximal convoluted tubule (PCT), exits the Bowman's capsule as a winding tube in the renal cortex. The wall of the PCT consists of cuboidal cells containing numerous mitochondria and bearing a brush border of dense microvilli that face the lumen (interior cavity). The high-energy yield and large surface area of these cells support their functions of reabsorption and secretion.

The middle of the tubule, the loop of Henle, is shaped like a hairpin and consists of a descending limb that drops into the renal medulla and an ascending limb that rises back into the renal cortex. As the loop descends, the tubule suddenly narrows, forming the thin segment of the loop. The loop subsequently widens in the ascending limb, forming the thick segment of the loop. Cells of the loop of Henle vary from simple squamous epithelium (descending limb and thin segment of ascending limb) to cuboidal and low columnar epithelium (thick segment of ascending limb) and almost entirely lack microvilli.

The final section, the distal convoluted tubule (DCT), coils within the renal cortex and empties into the collecting duct. Cells here are cuboidal with few microvilli.

Renal tubules of neighboring nephrons empty urine into a single collecting duct. Here and in the final portions of the DCT, two types of cells are present. The more numerous principal cells respond to the hormones aldosterone and antidiuretic hormone (ADH), and intercalated cells secrete H^+.

Various collecting ducts within the medullary pyramids merge to form papillary ducts, which drain into the minor calyces of the renal pelvis through the medullary papillae. Urine collects in the renal pelvis and drains out of the kidney through the ureter.

The efferent arteriole that carries blood away from the glomerulus branches to form peritubular capillaries. These capillaries weave around the portions of the renal tubule that lie in the renal cortex. In portions of the loop of Henle that descend deep into the renal medulla, the capillaries form loops, called vasa recta, that cross between the ascending and descending limbs. The peritubular capillaries collect water and nutrients from the filtrate in the tubule. They also release substances that are secreted into the tubule to combine with the filtrate in the formation of urine. The capillaries ultimately merge into an interlobular vein, which transports blood out of the nephron. Note that the blood flow through the nephron actually passes through two separate capillary beds, the glomerulus and the capillary network surrounding the renal tubule.

There are two kinds of nephrons:

■ Cortical nephrons, representing 85 percent of the nephrons in the kidney, have loops of Henle that descend only slightly into the renal medulla (refer to Figure 19-1).

■ Juxtamedullary nephrons have long loops of Henle that descend deep into the renal medullar. Only juxtamedullary nephrons have vasa recta that traverse their loops of Henle (refer to Figure 19-1).

The juxtaglomerular apparatus (JGA) is an area of the nephron where the afferent arteriole and the initial portion of the distal convoluted tubule are in close contact. Here, specialized smooth muscle cells of the afferent arteriole, called granular juxtaglomerular (JG) cells, are mechanoreceptors that monitor blood pressure in the afferent arteriole. In the adjacent distal convoluted tubule, specialized cells, called macula densa, are chemoreceptors that monitor the concentration of Na^+ and Cl^- in the urine inside the tubule. Together, these cells help regulate blood pressure and the production of urine in the nephron.

The operation of the human nephron consists of three processes:

■ Glomerular filtration

■ Tubular reabsorption

■ Tubular secretion

These three processes, which determine the quantity and quality of the urine, are discussed in the following sections.

Glomerular filtration

When blood enters the glomerulus, water and solutes are forced into the Bowman's capsule. Passage of cells and certain molecules are restricted as follows.

■ The fenestrae (pores) of the capillary endothelium are large, permitting all components of blood plasma to pass except blood cells.

■ A basement membrane (consisting of extracellular material) that lies between the capillary endothelium and the visceral layer of the Bowman's capsule blocks the entrance of large proteins into the Bowman's capsule.

■ The filtration slits between the pedicels of the podocytes prevent the passage of medium-sized proteins into the Bowman's capsule.

The net filtration pressure (NFP) determines the quantity of filtrate that is forced into the Bowman's capsule. The NFP, estimated at about 10 mm Hg, is the sum of pressures that promote filtration less the sum of those that oppose filtration. The following contribute to the NFP:

- The glomerular hydrostatic pressure (blood pressure in the glomerulus) promotes filtration.

- The glomerular osmotic pressure inhibits filtration. This pressure is created as a result of the movement of water and solutes out of the glomerular capillaries, while proteins and blood cells remain. This decreases the concentration of water in the glomerulus and promotes the return of water to the glomerulus by osmosis.

- The capsular hydrostatic pressure inhibits filtration. This pressure develops as water collects in the Bowman's capsule. The more water in the capsule, the greater the pressure.

The glomerular filtration rate (GFR) is the rate at which filtrate collectively accumulates in the Bowman's capsule of each nephron. The GFR, about 125 ml/min (180 liters/day), is regulated by the following:

- Renal autoregulation is the ability of the kidney to maintain a constant GFR even when the body's blood pressure fluctuates. Autoregulation is accomplished by cells in the juxtaglomerular apparatus that decrease or increase secretion of a vasoconstrictor substance that dilates or constricts, respectively, the afferent arteriole.

- Neural regulation of GFR occurs when vasoconstrictor fibers of the sympathetic nervous system constrict afferent arterioles. Such stimulation may occur during exercise, stress, or other fight-or-flight conditions and results in a decrease in urine production.

- Hormonal control of GFR is accomplished by the reninangiotensinogen mechanism. When cells of the juxtaglomerular apparatus detect a decrease in blood pressure in the afferent arteriole or a decrease in solute (Na^+ and Cl^-) concentrations in the distal tubule, they secrete the enzyme rennin. Renin converts angiotensinogen (a plasma protein produced by the liver) to angiotensinogen I. Angiotensinogen I, in turn, is converted to angiotensinogen II by angiotensin converting enzyme (ACE), an enzyme produced principally by capillary endothelium in the lungs. Angiotensinogen II circulates in the blood and increases GFR by

  constricting blood vessels throughout the body, causing the blood pressure to rise,

  constricting the afferent arterioles, and

  stimulating the adrenal cortex to secrete aldosterone, a hormone that increases blood pressure by decreasing water output by the kidneys.

Tubular reabsorption

In healthy kidneys, nearly all of the desirable organic substances (proteins, amino acids, glucose) are reabsorbed by the cells that line the renal tube. These substances then move into the peritubular capillaries that surround the tubule. Most of the water (usually more than 99 percent of it) and many ions are reabsorbed as well, but the amounts are regulated so that blood volume, pressure, and ion concentration are maintained within required levels for homeostasis.

Reabsorbed substances move from the lumen of the renal tubule to the lumen of a peritubular capillary. Three membranes are traversed:

■ The luminal membrane, or the side of the tubule cells facing the tubule lumen

■ The basolateral membrane, or the side of the tubule cells facing the interstitial fluids

■ The endothelium of the capillaries

Tight junctions between tubule cells prevent substances from leaking out between the cells. Movement of substances out of the tubule, then, must occur through the cells, either by active transport (requiring ATP) or by passive transport processes. Once outside of the tubule and in the interstitial fluids, substances move into the peritubular capillaries or vasa recta by passive processes.

The reabsorption of most substances from the tubule to the interstitial fluids, requires a membrane-bound transport protein that carries these substances across the tubule cell membrane by active transport. When all of the available transport proteins are being used, the rate of reabsorption reaches a transport maximum (Tm), and substances that cannot be transported are lost in the urine.

The following mechanisms direct tubular reabsorption in the indicated regions:

■ Active transport of Na^+ (PCT, DCT, collecting duct). Because Na^+ concentration is low inside tubular cells, Na^+ enters the tubular cells (across the luminal membrane) by passive diffusion. At the other side of the tubule cells, the basolateral membrane bears proteins that function as sodium-potassium (Na^+-K^+) pumps. These pumps use ATP to simultaneously export Na^+ while importing K^+. Thus, Na^+ in the tubule cells is transported out of the cells and into the interstitial fluid by active transport. The Na^+ in the interstitial fluid then enters the capillaries by passive diffusion. (The K^+ that is transported into the cell leaks back passively into the interstitial fluid.)

■ Symporter transport (secondary active transport) or nutrients and ions (PCT, loop of Henle). Various nutrients, such as glucose and amino acids, and certain ions (K^+ and Cl^-) in the thick ascending limb of the loop of Henle are transported into the tubule cells by the action of Na^+ symporters. An Na^+ symporter is a transport protein that carries both Na^+ and another molecule, such as glucose, across a membrane in the same direction. Movement of glucose and other nutrients form the tubular lumen into the tubule cells occurs in this fashion. The process requires a low concentration of Na^+ inside the cells, a condition maintained by the Na^+-K^+ pump operating on the basolateral membranes of the tubule cells. The movement of nutrients into cells by this mechanism is referred to as secondary active transport, because the ATP-requiring mechanism is the Na^+-K^+ pump and not the symporter itself. Once inside the tubular cells, nutrients move into the interstitial fluid and into the capillaries by passive processes.

■ Passive transport of H_2O by osmosis (PCT, DCT). The buildup of Na^+ in the peritubular capillaries creates a concentration gradient across which water passively moves, from tubule to capillaries, by osmosis. Thus, the reabsorption of Na^+ by active transport generates the subsequent reabsorption of H_2O by passive transport, a process called obligatory H_2O reabsorption.

■ Passive transport of various solutes by diffusion (PCT, DCT, and collecting duct). As H_2O moves from the tubule to the capillaries, various solutes, such as K^+, Cl^-, $HCO_3.$, and urea become more concentrated in the tubule. As a result, these solutes follow the water, moving by diffusion out of the tubule and into capillaries where their concentrations are lower, a process called solvent drag. Also, the accumulation of the positively charged $Na+$ in the capillaries creates an electrical gradient that attracts (by diffusion) negatively charged ions ($Cl-$, $HCO_3.$).

■ H_2O and solute transport regulated by hormones (DCT and collecting duct). The permeability of the DCT and collecting duct and the resultant reabsorption of H_2O and Na^+ are controlled by two hormones:

Aldosterone increases the reabsorption of Na^+ and H_2O by stimulating an increase in the number of Na^+-K^+ pump proteins in the principal cells that line the DCT and collecting duct.

Antidiuretic hormone (ADH) increases H_2O reabsorption by stimulating an increase in the number of H_2O-channel proteins in the principal cells of the collecting duct.

Tubular secretion

In contrast to tubular reabsorption, which returns substances to the blood, tubular secretion removes substances from the blood and secretes them into the filtrate. Secreted substances include H^+, K^+, NH_{4+} (ammonium ion), creatinine (a waste product of muscle contraction), and various other substances (including penicillin and other drugs). Secretion occurs in portions of the PCT, DCT, and collecting duct.

■ Secretion of H^+. Because a decrease in H^+ causes a rise in pH (a decrease in acidity), H^+ secretion into the renal tubule is a mechanism for raising blood pH. Various acids produced by cellular metabolism accumulate in the blood and require that their presence be neutralized by removing H^+. In addition, CO_2, also a metabolic byproduct, combines with water (catalyzed by the enzyme carbonic anhydrase) to produce carbonic acid (H_2CO_3), which dissociates to produce H^+, as follows.

$$CO_2 + H_2O \rightleftharpoons H_2CO_3 \rightleftharpoons H^+ + HCO_3{}_-$$

This chemical reaction occurs in either direction (it is reversible) depending upon the concentration of the various reactants. As a result, if $HCO_3{}_-$ increases in the blood, it acts as a buffer of $H+$, combining with it (and effectively removing it) to produce CO_2 and H_2O. An increase in acid levels in the blood, then, can be compensated for either by removing H^+ from the blood or by adding $HCO_3{}_-$ to the blood to buffer H^+ — that is, to combine the with the H^+ and neutralize its effect. In fact, renal secretion of H^+ accomplishes an increase in pH by this latter method.

CO_2 in tubular cells of the collecting duct combines with H_2O to form H^+ and $HCO_3{}_-$. The CO_2 may originate in the tubular cells or it may enter these cells by diffusion from the renal tubule, interstitial fluids, or peritubular capillaries. In the tubule cell, Na^+-H^+ antiporters, enzymes that move transported substances in opposite directions, transport H^+ across the luminal membrane into the tubule while importing $Na+$. Inside the tubule, H^+ may combine with any several buffers that entered the tubule as filtrate ($HCO_3{}_-$, NH_3, or $HPO_{42}{}_-$). If $HCO_3{}_-$ is the buffer, then H_2CO_3 is formed, producing H_2O and CO_2. The CO_2 then enters the tubular cell where it can combine with H_2O again. If $H+$ combines with another buffer, it is excreted in the urine. Regardless of the fate of the $H+$ in the tubule, the $HCO_3{}_-$ produced in the first step is transported across the basolateral membrane by an HCO_3—$Cl-$ antiporter. The $HCO_3{}_-$ enters the peritubular capillaries where it combines with the H^+ in the blood and increases the blood pH. Note that the blood pH is increased by adding HCO_{3-} to the blood, not by removing H^+.

■ Secretion of NH_3. When amino acids are broken down, they produce toxic NH_3. The liver converts most NH_3 to urea, a less toxic substance. Both enter the filtrate during glomerular filtration and are excreted in the urine. However, when the blood is very acidic, the tubule cells break down the amino acid glutamate, producing NH_3 and HCO_{3-}. The NH_3 combines with H^+, forming $NH4^+$, which is transported across the luminal membrane by an Na^+ antiporter and excreted in the urine. The HCO_{3-} moves to the blood (as discussed above for H^+ secretion) and increases blood pH.

■ Secretion of K^+. Nearly all of the K^+ in filtrate is reabsorbed during tubular reabsorption. When reabsorbed quantities exceed body requirements, excess K^+ is secreted back into the filtrate in the collecting duct and final regions of the DCT. Because aldosterone stimulates an increase in Na^+-K^+ pumps, K^+ secretion (as well as Na^+ reabsorption) increases with aldosterone.

Regulation of Urine Concentration

The loop of Henle of juxtamedullary nephrons is the apparatus that allows the nephron to concentrate urine. The loop is a countercurrent multiplier system in which fluids move in opposite directions through side-by-side, semipermeable tubes. Substances are transported horizontally, by passive or active mechanisms, from one tube to the other. The movement of the transported substances up and down the tubes results in a higher concentration of substances at the bottom of the tubes than at the tops of the tubes. Details of the process follow and are also shown in Figure 19-2.

■ The descending limb of the loop of Henle is permeable to H_2O, so H_2O diffuses out into the surrounding fluids. Because the loop is impermeable to Na^+ and Cl^- and because these ions are not pumped out by active transport, Na^+ and Cl^- remain inside the loop.

■ As the fluid continues to travel down the descending limb of the loop, it becomes more and more concentrated, as water continues to diffuse out. Maximum concentration occurs at the bottom of the loop.

■ The ascending limb of the loop of Henle is impermeable to water, but Na^+ and Cl^- are pumped out into the surrounding fluids by active transport.

■ As fluid travels up the ascending limb, it becomes less and less concentrated because Na^+ and Cl^- are pumped out. At the top of the ascending limb, the fluid is only slightly less concentrated than at the top of the descending limb. In other words, there is little change in the concentration of the fluid in the tubule as a result of traversing the loop of Henle.

■ In the fluid surrounding the loop of Henle, however, a gradient of salt (Na^+, Cl^-) is established, increasing in concentration from the top to the bottom of the loop.

Figure 19-2 The loop is a countercurrent multiplier system in which fluids move in opposite directions through side-by-side, semipermeable tubes. This process regulates the concentration of urine.

Regulation of Urine Concentration

■ Fluid at the top of the collecting duct has a concentration of salts about equal to that at the beginning of the loop of Henle (some water is reabsorbed in the DCT). As the fluid descends the collecting duct, the fluid is exposed to the surrounding salt gradient established by the loop of Henle. Without ADP, the collecting duct is impermeable to H_2O. Two outcomes are possible:

> If water conservation is necessary, ADH stimulates the opening of water channels in the collecting duct, allowing H_2O to diffuse out of the duct and into the surrounding fluids. The result is concentrated urine (refer to Figure 19-2).

If water conservation is not necessary, ADH is not secreted and the duct remains impermeable to H_2O. The result is dilute urine.

■ The vasa recta delivers O_2 and nutrients to cells of the loop of Henle. The vasa recta, like other capillaries, is permeable to both H_2O and salts and could disrupt the salt gradient established by the loop of Henle. To avoid this, the vasa recta acts as a countercurrent multiplier system as well. As the vasa recta descends into the renal medulla, water diffuses out into the surrounding fluids, and salts diffuse in. When the vasa recta ascends, the reverse occurs. As a result, the concentration of salts in the vasa recta is always about the same as that in the surrounding fluids, and the salt gradient established by the loop of Henle remains in place.

Ureters

The ureters, one from each kidney, deliver urine to the bladder. The ureters enter through the back of the bladder, entering at an angle such that when the bladder fills, the ureter openings are forced closed. A cross section of the ureter reveals three layers of tissue:

■ An inner mucosa consists of transitional epithelium covered by a lamina propria of connective tissue. Mucus secretions protect the ureter tissues from the urine.

■ A middle muscularis layer consists of longitudinal and circular layers of smooth muscle fibers. The muscle fibers force urine forward by peristalsis.

■ The outer adventitia consists of areolar connective tissue containing nerves, blood vessels, and lymphatic vessels.

Urinary Bladder

The urinary bladder is a muscular sac for storing urine. The triangular base of the bladder, the trigone, is defined by the two ureters that deliver the urine and the one urethra that drains the urine. When empty, the bladder collapses, and folds (called rugae) from in the bladder wall. As it fills, the folds become distended and the bladder becomes spherical. The wall of the bladder consists of three layers similar to those of the urethra: the mucosa, the muscularis (here called the detrusor muscle), and the adventitia. Circular smooth muscle fibers around the urethra form the internal urethral sphincter.

Urethra

The urethra drains urine from the bladder to an exterior opening of the body, the external urethral orifice. In females, the urethra is about 3 to 4 cm. (1.5 in.) long and opens to the outside of the body between the vagina and the clitoris. In males, the urethra is about 15 to 20 cm (6 to 8 in.) long and passes through the prostrate gland, the urogenital diaphragm, and the penis. In these regions, the urethra is called the prostatic urethra, membraneous urethra, and spongy (penile) urethra, respectively. In both males and females, a skeletal muscle, the external urethral sphincter, surrounds the urethra as it passes through the urogenital diaphragm.

Micturition, or urination, is the process of releasing urine from the bladder into the urethra. When the bladder fills to about 200 ml to 300 ml, stretch receptors in the bladder trigger a reflex arc. The signal stimulates the spinal cord which responds with a parasympathetic impulse that relaxes the internal urethral sphincter and contracts the detrusor muscle. Urine does not flow, however, until a voluntary nerve impulse relaxes the skeletal muscle of the external urethral sphincter.

Chapter Checkout

Q&A

1. True or False. The kidney consists of over a billion individual filtering units called nephrons.

2. The urinary system helps maintain _____ by regulating water balance and by removing harmful substances from the blood.

3. Which of the following is not a process of the human nephron?
   a. Tubular secretion
   b. Homeostasis
   c. Glomerular filtration
   d. Tubular reabsorption

4. True or False. The ureters, one from each kidney, deliver urine to the bladder.

5. The _____ _____ is a muscular sac for storing urine.

Answers: 1. F 2. homeostasis 3. b 4. T 5. urinary bladder

Chapter 20

THE REPRODUCTIVE SYSTEM

Chapter Check-In

❑ Listing the main structures of the male and female reproductive systems

❑ Knowing the structures of a sperm cell

❑ Understanding the hormonal regulation of sperm production

❑ Describing the function of the mammary glands

❑ Comprehending the ovarian cycle and the menstrual cycle

This is the chapter about the birds and the bees, so to speak. A basic understanding of the reproductive organ and sex organs in both the male and female anatomy is critical to the understanding of the reproductive systems as a whole. You may recognize most of this information from your high school sex education class.

So What Is Reproduction?

Reproduction describes the production of eggs and sperm and the processes leading to fertilization. The reproductive system consists of the primary sex organs, or gonads, (testes in males and ovaries in females), which secrete hormones and produce gametes (sperm and eggs). Accessory reproductive organs include ducts, glands, and the external genitalia.

The Male Reproductive System

The **male reproductive system** consists of the following structures, shown in Figure 20-1:

- The scrotum is a sac consisting of skin and superficial fascia that hangs from the base of the penis. A vertical septum divides the scrotum into left and right compartments, each of which encloses a testis. The external scrotum positions the testes outside the body in an environment about 3° C below that of the body cavity, a condition necessary for the development and storage of sperm. The following two muscles help maintain this temperature:

   The dartos muscle is located in the superficial fascia of the scrotum and septum. Contraction of this smooth muscle creates wrinkles in the scrotum skin. The wrinkling thickens the skin, reducing heat loss when external temperatures are too cold.

   The cremaster muscles extend from the internal oblique muscle to the scrotum. Contraction of these skeletal muscles lifts the scrotum closer to the body when external temperatures are too cold.

- Each of the two testes (singular, testis) consists of the following structures:

   The tunica vaginalis is a two-layer outer serous membrane surrounding each testis.

   The tunica albuginea lies inside the tunica vaginalis and protrudes inward, dividing each testis into compartments called lobules.

   One to four tightly coiled tubes, the seminiferous tubules, lie inside each lobule. The seminiferous tubules are the sites of the sperm production (spermatogenesis). The tubule is lined with spermatogenic cells, cells that form sperm, and sustentacular cells (Sertoli cells), cells that support the developing sperm. The coiled seminiferous tubules inside each lobule unite to form a straight tube, the tubulus rectus.

   The rete testis is a network of tubes formed by the merging of the tubules recti from each lobule.

   The efferent ducts transport sperm out of the testis (from the rete testis) to the epididymis.

   Interstitial cells (Leydig cells) surrounding the seminiferous tubules secrete testosterone and other androgen hormones.

Figure 20-1 View of the male reproductive system.

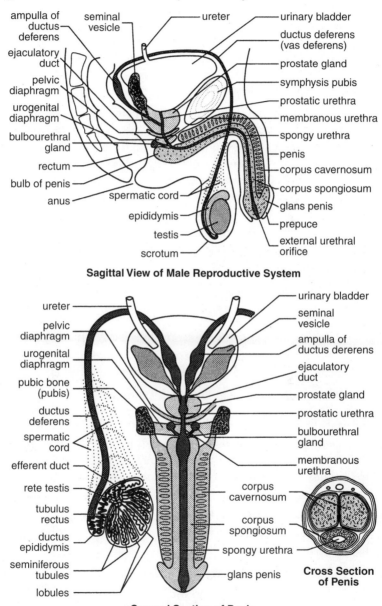

Sagittal View of Male Reproductive System

Coronal Section of Penis

Cross Section of Penis

- The epididymis is a comma-shaped organ that lies adjacent to each testis. Each of the two epididymides contains a tightly coiled tube, the ductus epididymis. Here, sperm complete their maturation and are stored until ejaculation. During ejaculation, smooth muscles encircling the epididymis contract, forcing mature sperm into the next tube, the ductus deferens. The walls of the ductus epididymis contain microvilli called stereocilia that nourish sperm.

- The ductus deferens (vas deferens) is the tube through which sperm travel when they leave the epididymis. Each of the two tubes enter the abdominal cavity, passes around the bladder (refer to Figure 20-1), and together with the duct from the seminal vesicle, joins the ejaculatory duct. Before entering the ejaculatory duct, the ductus deferens enlarges, forming a region called the ampulla. Sperm are stored in the ductus deferens until peristaltic contractions of the smooth muscles surrounding the ductus force sperm forward during ejaculation.

- The ejaculatory ducts are short tubes that connect each ductus deferens to the urethra.

- The urethra is the passageway for urine and semen (sperm and associated secretions). Three regions of the urethra are distinguished:

    The prostatic urethra passes through the prostate gland.

    The membranous urethra passes through the urogenital diaphragm (muscles associated with the pelvic region).

    The spongy (penile) urethra passes through the penis.

    The urethra ends at the external urethral orifice.

- The spermatic cord contains blood vessels, lymphatic vessels, nerves, the ductus deferens, and the cremaster muscle. It connects each testis to the body cavity, entering the abdominal wall through the inguinal canal.

- The accessory sex glands are glands that secrete substances into the passageways that transport sperm. These substances contribute to the liquid portion of the semen.

    The seminal vesicles secrete into the vas deferens an alkaline fluid (which neutralizes the acid in the vagina), fructose (which provides energy for the sperm), and prostaglandins (which increase sperm viability and stimulate female uterine contractions that help sperm move into the uterus).

The prostate gland secretes a milky, slightly acidic fluid into the urethra. Various substances in the fluid increase sperm mobility and viability.

The bulbourethral (Cowper's) glands secrete an alkaline fluid into the spongy urethra. The fluid neutralizes acidic urine in the urethra before ejaculation occurs.

■ The penis is a cylindrical organ that passes urine and delivers sperm. It consists of a root that attaches the penis to the perineum, a body (shaft) that makes up the bulk of the penis, and the glans penis, the enlarged end of the body. The glans penis is covered by a prepuce (foreskin), which may be surgically removed by a procedure called circumcision. Internally, the penis consists of three cylindrical masses of tissue, each of which is surrounded by a thin layer of fibrous tissue, the tunic albuginea. The three cylindrical masses, which function as erectile bodies, are as follows:

Two corpora cavernosa fill most of the volume of the penis. Their bases, called the crura (singular, crus) of the penis, attach to the urogenital diaphragm.

A single corpus spongiosum encloses the urethra and expands at the end to form the glans penis. The bulb of the penis, an enlargement at the base of the corpus spongiosum, attaches to the urogenital diaphragm.

During erection, parasympathetic neurons stimulate dilation of the arteries that deliver blood to the corpus cavernosa and spongiosum. As a result, blood collects in these blood vessels and causes the penis to enlarge and stiffen. Ejaculation occurs when sympathetic neurons stimulate the discharge of sperm and supporting fluids from their various sources. During ejaculation, the sphincter muscle at the base of the urinary bladder constricts, preventing the passage of urine.

Spermatogenesis

The cells that line the walls of the seminiferous tubules are collectively called spermatogenic cells. Those cells nearest the basement membrane are called spermatogonia. These cells are stem cells—that is, they are capable of continuous division and remain undifferentiated, never maturing into specialized cells. Extending from the spermatogonia toward the lumen of the tubule are cells at various levels of maturity, with the most mature cells—the sperm—facing the lumen.

Spermatogenesis begins at puberty within the seminiferous tubules of the testes. The spermatogonia, each of which contains 46 chromosomes, divide by mitosis repeatedly to produce primary spermatocytes (still diploid cells with 46 chromosomes each). The primary spermatocytes begin meiosis. During the first meiotic division (meiosis I, or the reduction division), each primary spermatocyte divides into two secondary spermatocytes, each with 23 chromosomes (haploid cells). During the second meiotic division (meiosis II, or the equatorial division), each secondary spermatocyte divides again, producing a total of four spermatids. Each spermatid still contains 23 chromosomes, but these chromosomes consist of only one chromatid (rather than the normal two chromatids).

Spermiogenesis describes the development of spermatids into mature sperm (sperm cells, or spermatozoa). At the end of this process, each sperm cell bears the following structures:

■ The head of the sperm contains the haploid nucleus with 23 chromosomes. At the tip of the sperm head is the acrosome, a lysosome containing enzymes which are used to penetrate the egg. The acrosome originates from Golgi body vesicles that fuse to form a single lysosome.

■ The midpiece is the first part of the tail. Mitochondria spiral around the midpiece and produce energy (ATP) used to generate the whip-like movements of the tail that propel the sperm.

■ The tail is a flagellum consists of the typical 9 + 2 microtubule array.

Hormonal regulation of spermatogenesis

The production of sperm is regulated by hormones, as shown in Figure 20-2:

■ The hypothalamus begins secreting gonadotropin-releasing hormone (GnRH) at puberty.

■ GnRH stimulates the anterior pituitary to secrete follicle-stimulating hormone (FSH) and luteinizing hormone (LH).

■ LH stimulates the interstitial cells in the testes to produce testosterone and other male sex hormones (androgens). (In males, LH is also called interstitial cell stimulating hormone, or ICSH.)

Figure 20-2 Processes of hormone regulation in the male and female reproductive systems.

(a) Hormone Regulation in the Male Reproductive System

(b) Hormone Regulation in the Female Reproductive System

■ Testosterone produces the following effects:

Testosterone stimulates the final stages of sperm development in the nearby seminiferous tubules. It accumulates in these tissues because testosterone and FSH act together to stimulate sustentacular cells to release androgen-binding protein (ABP). ABP holds testosterone in these cells.

Testosterone entering the blood circulates throughout the body where it stimulates activity in the prostate gland, seminal vesicles, and various other target tissues.

Testosterone and other androgens stimulate the development of secondary sex characteristics, those characteristics not directly involved in reproduction. These include the distribution of muscle and fat typical in adult males, various body hair (facial and pubic hair, for example), and deepening of the voice.

Levels of testosterone are regulated by a negative-feedback mechanism with the hypothalamus. When the hypothalamus detects excessive amounts of testosterone in the blood, it reduces its secretion of GnRH. In response, the anterior pituitary reduces its production of LH and FSH, which results in a decrease in the production of testosterone by interstitial cells. GnRH secretion is also inhibited by inhibin, a hormone secreted by sustentacular cells in response to excessive levels of sperm production.

The Female Reproduction System

The **female reproductive system** consists of the following structures, as shown in Figure 20-3:

■ The ovary is the organ that produces ova (singular, ovum), or eggs. The two ovaries present in each female are held in place by the following ligaments:

The mesovarium is a fold of peritoneum that holds the ovary in place.

The suspensory ligament anchors the upper region of the ovary to the pelvic wall. Attached to this ligament are blood vessels and nerves, which enter the ovary at the hilus.

The broad ligament is a section of the peritoneum that drapes over the ovaries and uterus. It includes both the mesovarium and suspensory ligament.

The ovarian ligament anchors the lower end of the ovary to the uterus.

Figure 20-3   View of the female reproductive system.

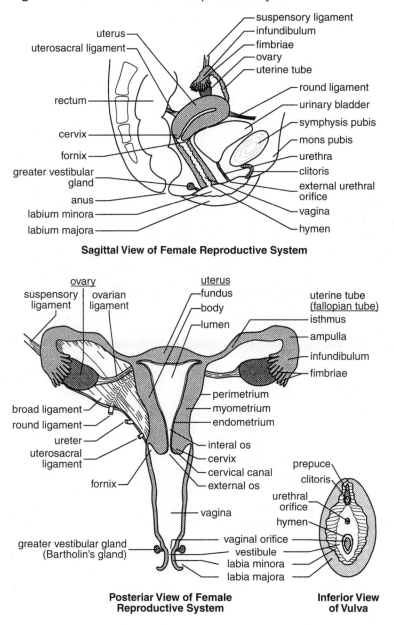

Sagittal View of Female Reproductive System

Posteriar View of Female
Reproductive System

Inferior View
of Vulva

The following two tissues cover the outside of the ovary:

> The germinal epithelium is an outer layer of simple epithelium.
>
> The tunica albuginea is a fibrous layer inside the germinal epithelium.

The inside of the ovary, or stroma, is divided into two indistinct regions, the outer cortex and the inner medulla. Embedded in the cortex are saclike bodies called ovarian follicles. Each ovarian follicle consists of an immature oocyte (egg) surrounded by one or more layers of cells that nourish the oocyte as it matures. The surrounding cells are called follicular cells, if they make up a single layer, or granulosa cells, if more than one layer is present.

■ The uterine tubes (fallopian tubes, or oviducts) transport the secondary oocytes away from the ovary and toward the uterus. The following regions characterize each of the two uterine tubes (one for each ovary):

> The infundibulum is a funnel-shaped region of the uterine tube that bears fingerlike projections called fimbriae. Pulsating cilia on the fimbriae draw the secondary oocyte into the uterine tube.
>
> The ampulla is the widest and longest region of the uterine tube. Fertilization of the oocyte by a sperm usually occurs here.
>
> The isthmus is a narrow region of the uterine tube whose terminus enters the uterus.

■ The wall of the uterine tube consists of the following three layers:

> The serosa, a serous membrane, lines the outside of the uterine tube.
>
> The middle muscularis consists of two layers of smooth muscle that generate peristaltic contractions that help propel the oocyte forward.
>
> The inner mucosa consists of ciliated columnar epithelial cells that help propel the oocyte forward and secretory cells that lubricate the tube and nourish the oocyte.

■ The uterus (womb) is a hollow organ within which fetal development occurs. The uterus is characterized by the following regions:

> The fundus is the upper region where the uterine ducts join the uterus.

The body is the major, central portion of the uterus.

The isthmus is the lower, narrow portion of the uterus.

The cervix is a narrow region at the bottom of ht uterus that leads to the vagina. The inside of the cervix, or cervical canal, opens to the uterus above through the internal os and to the vagina below through the external os. Cervical mucus secreted by the mucosa layer of the cervical canal serves to protect against bacteria entering the uterus from the vagina. If an oocyte is available for fertilization, the mucus is thin and slightly alkaline, attributes that promote the passage of sperm. At other times, the mucus is viscous and impedes the passage of sperm.

■ The uterus is held in place by the following ligaments:

The broad ligaments

The uterosacral ligaments

The round ligaments

The cardinal (lateral cervical) ligaments

■ The wall of the uterus consists of the following three layers:

The perimetrium is a serous membrane that lines the outside of the uterus.

The myometrium consists of several layers of smooth muscle and imparts the bulk of the uterine wall. Contractions of these muscles during childbirth help force the fetus out of the uterus.

The endometrium is the highly vascularized mucosa that lines the inside of the uterus. If an oocyte has been fertilized by a sperm, the zygote (the fertilized egg) implants on this tissue. The endometrium itself consists of two layers. The stratum functionalis (functional layer) is the innermost layer (facing the uterine lumen) and is shed during menstruation. The outermost stratum basalis (basal layer) is permanent and generates each new stratum functionalis.

■ The vagina (birth canal) serves both as the passageway for a newborn infant and as a depository for semen during sexual intercourse. The upper region of the vagina surrounds the protruding cervix, creating a recess called the fornix. The lower region of the vagina opens to the outside at the vaginal orifice. A thin membrane called the hymen may cover the orifice. The vaginal wall consists of the following layers:

The outer adventitia holds the vagina in position.

The middle muscularis consists of two layers of smooth muscle that permit expansion of the vagina during childbirth and when the penis is inserted.

The inner mucosa has no glands. But bacterial action on glycogen stored in these cells produces an acid solution that lubricates the vagina and protects it against microbial infection. The acidic environment is also inhospitable to sperm. The mucosa bears transverse ridges called rugae.

■ The vulvae (pudendum) make up the external genitalia. The following structures are included:

The mons pubis is a region of adipose tissue above the vagina that is covered with hair.

The labia majora are two folds of adipose tissue that border each side of the vagina. Hair and sebaceous and sudoriferous glands are present. Developmentally, the labia majora are analogous to the male scrotum.

The labia minora are smaller folds of skin that lie inside the labia majora.

The vestibule is the recess formed by the labia minora. It encloses the vaginal orifice, the urethral opening, and ducts from the greater vestibular glands whose mucus secretions lubricate the vestibule.

The clitoris is a small mass of erectile and nervous tissue located above the vestibule. Extensions of the labia minora join to form the prepuce of the clitoris, a fold of skin covering the clitoris.

Mammary glands

The **mammary glands** are sudoriferous (sweat) glands specialized for the production of milk. The milk-producing secretory cells form walls of bulb-shaped chambers called alveoli that join together with ducts, in grapelike fashion, to form clusters called lobules. Numerous lobules assemble to form a lobe. Each breast contains a single mammary gland consisting of 15 to 20 of these lobes. Lactiferous ducts leading away from the lobes widen into lactiferous sinuses that serve as temporary reservoirs for milk. The ducts narrow again as they lead through a protruding nipple. The nipple, whose texture is made coarse by the presence of sebaceous glands, is surrounded by a ring of pigmented skin called the areola. Contraction of myoepithelial cells surrounding the alveoli force milk toward the nipples.

The breasts begin to enlarge in females at the onset of puberty. Proliferating adipose (fat) tissue expands the breast, while suspensory ligaments (Cooper's ligaments) attached to the underlying fascia provide support. In nonpregnant females (and males), the glands and ducts are not fully developed.

During pregnancy, estrogen and progesterone stimulate extensive development of the mammary glands and associated ducts. After childbirth, various hormones, especially prolactin from the anterior pituitary, initiate lactation, or milk production. When neurons are stimulated by the sucking of an infant, nerve impulses activate the posterior pituitary to secrete oxytocin, which, in turn, stimulates contraction of the myoepithelial cells surrounding the alveoli. Milk is then forced toward the nipple (the letdown reflex).

Oogenesis

Oogenesis consists of the meiotic cell divisions that lead to the production of ova (eggs) in females. The process begins during fetal development with the fetal ovary. Diploid cells called oogonia divide by mitosis to produce primary oocytes (still diploid with 46 chromosomes). Each primary oocyte is encircled by one or more layers of cells. The oocyte and encircling cells together are called an ovarian follicle. The primary oocytes (within their follicles) begin meiosis, but division progresses only to prophase I. They remain at this stage until puberty.

The following stages in the development of an ovarian follicle are observed:

- The primordial follicle, the initial fetal state of the follicle, encircles the oocyte with a single layer of cells called follicular cells.

- The primary follicle, the next stage of follicular development, possesses two or more layers of encircling cells now called granulosa cells.

- The secondary follicle is distinguished by the presence of the antrum, a fluid-filled, central cavity.

- In a mature (vesicular, or Graafian) follicle, the primary oocyte has completed meiosis I. It is the stage of follicular development that precedes ejection of the oocyte from the ovary (ovulation). The following features are observed:

     The zona pellucida, a clear layer of glycoprotein, surrounds the oocyte.

The corona radiata, a ring of granulosa cells, encircles the zona pellucida.

The theca folliculi, the ovarian cells immediately surrounding the outer layers of granulosa cells, differentiate into an internal layer (facing the follicle) of secretory cells, the theca interna, and an external layer of cells, the theca interna, and an external layer of connective tissue, the theca externa.

■ The corpus luteum is the remains of the follicle following ovulation. It remains functional, producing estrogen, progesterone, and inhibin, until it finally degenerates.

During each menstrual cycle, one primary oocyte, enclosed in its follicle, resumes meiosis I to produce two daughter cells (each haploid with 23 chromosomes). One daughter cell, the secondary oocyte, contains most of the cytoplasm, ensuring that adequate amounts of stored food, as well as mitochondria, ribosomes, and other cytoplasmic organelles, will be available for the developing embryo. The other daughter cell, a first polar body, is much smaller, and contains little cytoplasm and few, if any, organelles. The secondary oocyte then begins meiosis II (equatorial division) but again stops at prophase (this time prophase II). The first polar body may also begin meiosis II, but it will eventually degenerate.

Ovulation occurs when a secondary oocyte and its first polar body, surrounded by the zona pellucida and corona radiata, rupture from their mature follicle and are expelled from the surface of the ovary. The oocyte is then swept up into the uterine (fallopian) tube and advances toward the uterus. If a sperm cell penetrates the corona radiata and zona pellucida and enters the secondary oocytes, meiosis II resumes in the secondary oocytes, producing an ovum and a second polar body. If a first polar body is present, it, too, may resume meiosis II, producing daughter polar bodies. Fertilization occurs when the nuclei of the sperm cell and ovum unite, forming a zygote (fertilized egg). Any polar bodies present ultimately degenerate.

Hormonal regulation of oogenesis and the menstrual cycle

The human female reproductive cycle is characterized by events in the ovary (ovarian cycle) and the uterus (menstrual cycle). The purpose of these cycles is to produce an egg and to prepare the uterus for the implantation of the egg, should it become fertilized. The ovarian cycle consists of three phases:

■ The follicular phase describes the development of the follicle, the meiotic stages of division leading to the formation of the secondary oocytes, and the secretion of estrogen from the follicle.

■ Ovulation, occurring at midcycle, is the ejection of the egg from the ovary.

■ The luteal phase describes the secretion of estrogen and progesterone from the corps luteum (previously the follicle) after ovulation.

The menstrual (uterine) cycle consists of three phases:

■ The proliferative phase describes the thickening of the endometrium of the uterus, replacing tissues that were lost during the previous menstrual cycle.

■ The secretory phase follows ovulation and describes further thickening and vascularization of the endometrium in preparation for the implantation of a fertilized egg.

■ The menstrual phase (menstruation, menses) describes the shedding of the endometrium when implantation does not occur.

The activities of the ovary and the uterus are coordinated by negative and positive feedback responses involving gonadotrpin releasing hormone (GnRH) from the hypothalamus, follicle stimulating hormone (FSH) and luteinizing hormone (LH) from the anterior pituitary, and the hormones estrogen and progesterone from the follicle and corpus luteum. A description of the events follows (see Figures 20-2 and 20-4).

■ The hypothalamus and anterior pituitary initiate the reproductive cycle. The hypothalamus monitors the levels of estrogen and progesterone in the blood. In a negative-feedback fashion, low levels of these hormones stimulate the hypothalamus to secrete GnRH, which, in turn, stimulates the anterior pituitary to secrete FSH and LH.

■ The follicle develops. FSH stimulates the development of the follicle from primary through mature stages.

■ The follicle secretes estrogen. LH stimulates the cells of the theca interna and the granulosa cells of the follicle to secrete estrogen. Inhibin is also secreted by the granulosa cells.

Figure 20-4 The female reproductive cycle broken down by days.

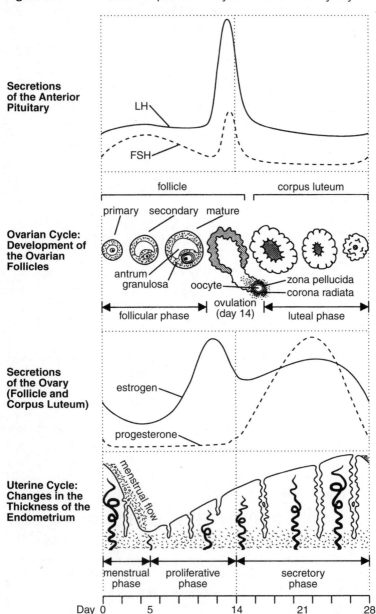

■ Ovulation occurs. Positive feedback from rising levels of estrogen stimulate the anterior pituitary (through GnRH from the hypothalamus) to produce a sudden midcycle surge of LH. This high level of LH stimulates meiosis in the primary oocyte to progress toward prophase II and triggers ovulation.

■ The corpus luteum secretes estrogen and progesterone. After ovulation, the follicle, now transformed into the corpus luteum, continues to develop under the influence of LH and secretes both estrogen and progesterone.

■ The endometrium thickens. Estrogen and progesterone stimulate the development of the endometrium, the inside lining of the uterus. It thickens with nutrient-rich tissue and blood vessels in preparation for the implantation of a fertilized egg.

■ The hypothalamus and anterior pituitary terminate the reproductive cycle. Negative feedback from the high levels of estrogen and progesterone cause the anterior pituitary (through the hypothalamus) to abate the production of FSH and LH. Inhibin also suppresses production of FSH and LH.

■ The endometrium either disintegrates or is maintained, depending upon whether implantation of the fertilized egg occurs, as follows:

> Implantation does not occur. In the absence of FSH and LH, the corpus luteum deteriorates. As a result, estrogen and progesterone production stops. Without estrogen and progesterone, growth of the endometrium is no longer supported, and it disintegrates, sloughing off during menstruation.

> Implantation occurs. The implanted embryo secretes human chorionic gonadotropin (hCG) to sustain the corpus luteum. The corpus luteum continues to produce estrogen and progesterone, maintaining the endometrium. (Pregnancy tests check for the presence of hCG in the urine.)

In addition to influencing the reproductive cycle, estrogen stimulates the development of secondary sex characteristics in females. These include the distribution of adipose tissue (to the breasts, hips, and mons pubis), bone development leading to a broadening of the pelvis, changes in voice quality, and growth of various body hair.

Chapter Checkout

Q&A

1. The development of spermatids into mature sperm (sperm cells, or spermatozoa) is called _____.

2. Which of the following is not true about testosterone?
   a. It stimulates the final stages of sperm development.
   b. It stimulates activity in the prostate gland, seminal vesicles, and various other target tissues.
   c. It neutralizes acid in the vagina
   d. It stimulates the development of secondary sex characteristics.

3. The _____ _____ are sudoriferous (sweat) glands specialized for the production of milk.

4. True or False. The menstrual (uterine cycle) consists of three phases: proliferative phase, the secretory phase, and the luteal phase.

5. True or False. Accessory reproductive organs include ducts, glands, and the external genitalia.

Answers: 1. spermiogenesis 2. c 3. mammary glands 4. F 5. T

CQR REVIEW

Use this CQR Review to practice what you've learned in this book. After you work through the review questions, you're well on your way to achieving your goal of learning about Anatomy and Physiology.

Chapter 1

1. Which of the following is not a chemical bond?

   a. Ionic
   b. Oxygen
   c. Covalent
   d. Hydrogen

2. True or False: Inorganic compounds are typically compounds without carbon atoms.

3. A _____ is the simplest kind of carbohydrate.

4. DNA stands for _____ _____.

5. True or False. Enzymes act as catalysts for metabolic reactions.

Chapter 2

6. Which of the following is not a function of the plasma membrane?

   a. Bounds the cell
   b. Controls the movement of materials into and out of the cell
   c. Encloses the nucleus and cytoplasm
   d. Absorbs extra fzluid into the cell

7. True or False. Organelles are bodies within the cytoplasm that serve to physically separate the various metabolic reactions that occur outside the cells.

8. What are the three kinds of cell junctions?

9. What are the differences in mitosis and meiosis?

Chapter 3

10. List the four categories of tissue.

11. True or False. Proteoglycans hold the connective tissue together.

12. Which of the following is not part of a neuron?

    a. Cell body
    b. Axon
    c. Dendrites
    d. Cutaneous membrane

13. _____ , like skeletal muscle, is striated.

Chapter 4

14. True or False. The skin consists of one layer, the dermis.

15. _____ of vitamin D occurs in the skin.

Chapter 5

16. The skeletal system consists of bones, _____ , and membranes that line the bones.

17. Which of the following is not a function of bone?

    a. Protection
    b. Thermoregulation
    c. Blood cell production
    d. Movement

18. True or False. Remodeling is the process of creating new bone and removing old bone.

Chapter 6

19. How many bones are in the human body?

20. Which bone is isolated from all other bones?

    a. Frontal bone
    b. Atlas
    c. Hyoid bone
    d. Axis

21. True or False. The coccyx, formed by three fused vertebrae, is a small triangle-shaped bone that attaches to the bottom of the sacrum.

Chapter 7

22. What are the three structural classes of joints?

23. What are the three functional classes of joints?

Chapter 8

24. Name the three types of muscle.

25. True or False. The endomysium is the connective tissue that surrounds each muscle fiber (cell).

26. A gap called a _____ or _____ separates the neuron from a muscle cell or another neuron.

Chapter 9

27. True or False. A skeletal muscle may attach a bone to another bone or a bone to another structure, such as skin.

28. Muscle fibers are grouped into _____, which are, in turn, grouped together to form a muscle.

29. Name four of the seven characteristics by which skeletal muscles are named.

Chapter 10

30. True or False. A nerve tract is a bundle of nerve fibers in the peripheral nervous system.

31. Cells that support and protect neurons are called _____.

32. Which of the following is not an event that characterizes the transmission of a nerve impulse?

    a. Action potential
    b. Repolarization
    c. Degradation potential
    d. Resting potential

33. In electrical synapses, the action potential travels along the membranes of _____ _____, small tubes of cytoplasm that allow the transfer of ions between adjacent cells.

Chapter 11

34. Name the two parts of the nervous system.

35. True or False. The somatic nervous system controls the activities of organs, glands, and various involuntary muscles, such as cardiac and smooth muscles.

36. The _____ nervous system activates tranquil functions, such as stimulating the secretion of saliva or digestive enzymes into the stomach.

37. True or False. The meninges are protective coverings of the brain and spinal cord.

Chapter 12

38. Which of the following is not a classification for sensory receptors?

    a. Function
    b. Receptor complexity
    c. Location
    d. Type of stimulus detected

39. True or False. The eye is supported by accessory organs.

40. The process of _____ involves converting light energy to chemical energy.

Chapter 13

41. The endocrine system produces _____ that are instrumental in maintaining homeostasis and regulating reproduction and development.

42. Most hormone production is regulated by a _____ feedback system.

Chapter 14

43. Name the three main functions of the cardiovascular system.

44. Which of the following is not true of an erythrocyte (red blood cell)?

    a. Shaped like flattened donuts
    b. Contains both protein and nonprotein groups
    c. Has a nucleus
    d. Transports oxygen and carbon dioxide in the blood

45. What are two important group proteins that stimulate the immune response?

46. True or False. The two circuits of blood circulation through the body are pulmonary circulation and myocardial circulation.

Chapter 15

47. True or False. The lymphatic system collects excess fluids and plasma proteins from surrounding tissues and returns them to the blood circulation.

48. _____ are white blood cells that provide an immune response that attacks specific kinds of nonself cells and foreign substances.

49. Which of the following is not a function of the lymph nodes?

   a. Filter the lymph
   b. Produce antibodies
   c. Destroy bacteria
   d. Produce toxins

Chapter 16

50. _____ is a physical and hostile barrier covered with oily and acidic secretions from sebaceous and sweat glands, respectively.

51. Which of the following is not a nonspecific barrier of the immune system?

   a. Cilia
   b. Gastric juice
   c. Phagocytes
   d. Skin

52. The primary agents of the immune response are _____, white blood cells that originate in the bone marrow.

Chapter 17

53. What is the function of the respiratory system?

54. Which of the following is not a process of the respiratory system?

   a. Internal respiration
   b. Absorb excess fluid
   c. Gas transport
   d. Pulmonary ventilation

55. Name the two ways that oxygen is transported in the blood.

Chapter 18

56. The function of the digestive system is _____ and _____.

57. List the seven processes for the treatment of food in the digestive system.

58. True or False. One of the functions of the liver is to regulate blood glucose levels.

Chapter 19

59. What is the main function of the urinary system?

60. The kidney consists of over a million individual filtering units called _____.

61. True or False. The ureters drain urine from the bladder an exterior opening of the body.

Chapter 20

62. The reproductive system consists of the _____ _____ _____, or gonads, which secrete hormones and produce gametes.

63. Which of the following is not a structure of the female reproductive system?

   a. Ovary
   b. Urethra
   c. Uterus
   d. Vulvae

64. _____ describes the development of spermatids into mature sperm.

65. True or False. The mammary glands are sudoriferous (sweat glands specialized for the production of lactic acid.

Answers: 1. b, **2.** T, **3.** monosaccharide, **4.** deoxyribonucleic acid, **5.** T, **6.** d, **7.** F, **8.** desmosomes, tight junctions, and gap junctions, **9.** Provide your own answer, **10.** epithelial tissue, connective tissue, nervous tissue, and muscle tissue, **11.** F, **12.** d, **13.** cardiac muscle, **14.** F, **15.** metabolism, **16.** cartilage, **17.** b, **18.** T, **19.** 206, **20.** c, **21.** F, **22.** fibrous, cartilaginous, and synovial, **23.** synarthrosis, amphiarthrosis, and diarthrosis, **24.** skeletal muscles, cardiac muscle, and smooth muscle, **25.** T, **26.** synapse or synaptic cleft, **27.** T, **38.** fascicles, **29.** number of origins, location of origin or insertion, location, shape, direction of muscle fibers, size, and action, **30.** F, **31.** neuroglia, **32.** c, **33.** gap junctions, **34.** central and peripheral, **35.** F, **36.** parasympathetic, **37.** T, **38.** a, **39.** T, **40.** sight, **41.** hormones, **42.** negative, **43.** transport, protection, and regulation, **44.** c, **45.** ABO blood group and Rh blood group, **46.** F, **47.** T, **48.** lymphocytes, **49.** d, **50.** skin, **51.** c, **52.** lymphocytes, **53.** to deliver air to the lungs, **54.** b, **55.** carried in the plasma as gas and carried in the blood bound to a protein, hemoglobin in red blood cells, **56.** digestion, absorption, **57.** ingestion, propulsion, secretion, mechanical digestion, chemical digestion, absorption, and defecation, **58.** T **59.** The urinary system helps maintain homeostasis by regulating water balance and by removing harmful substances from the blood, **60.** nephrons, **61.** F, **62.** primary sex organs, **63.** b, **64.** spermatogenesis, **65.** F

CQR RESOURCE CENTER

CQR Resource Center offers the best resources available in print and online to help you study and review the core concepts of Anatomy and Physiology. You can find additional resources, plus study tips and tools to help test your knowledge, at www.cliffsnotes.com.

Books

This CliffsQuickReview book is one of many great books about Anatomy and Physiology. If you want some additional resources, check out these other publications:

Grays Anatomy, by Henry Gray, covers almost anything you want to know about the human body. From its simplest to its most complex parts, you can find diagrams and illustrations accompanied by clear text that gives you the information you need quickly. Running Press Book Publishers $18.98.

The Anatomy Coloring Book, by Wynn Kapit, teaches concepts, structures, and systems of the body through the coloring process. HarperCollins College Publishers $17.99.

A Pictoral Handbook of Anatomy and Physiology, by James Bevan and Richard Bayliss, contains colorful graphics and straightforward text, minus the clinical jargon. Written for medical students to general readers. Barnes & Noble Books, $8.98.

Essentials of Anatomy & Physiology, 2nd Edition, by Frederic H. Martini and Edwin Bartholomew, contains hundreds of color photographs, drawings, and charts, introducing the human body system by system. Prentice Hall $79.75.

Grays Anatomy: A Fact-Filled Coloring Book, by Fred Stark, contains sixty detailed drawings designed for coloring, enhanced by detailed text describing each illustration. Running Press Book Publishers $8.95.

Fundamentals of Anatomy and Physiology, by Roberta M. Meehan, gives you a detailed look at each system of the body. Prentice Hall, $46.00.

Wiley also has three Web sites that you can visit to read about all the books we publish:

- www.cliffsnotes.com
- www.dummies.com
- www.wiley.com

Internet

Visit the following Web sites for more information about Anatomy and Physiology:

3Danatomy.com, www.3danatomy.com, gives you realistic images and animations. You can purchase 3D anatomy software as well as get free downloads.

adam.com, www.adam.com, features adam online anatomy which has been exhaustively reviewed by panels of leading anatomists, is recognized as the standard anatomical database in computer-based medical education worldwide. It also features Ipweb (Interactive Physiology), which offers an exciting online exploration of physiology and anatomy through engaging content and imagery. IPweb uses content-rich tutorials, Shockwave animations, quizzes and tests to explain physiological concepts and processes that are challenging to teach and learn. Subscriptions are available from 6 to 12 months and prices range from $19.95 to $84.95 depending on your package choice and length of subscription.

The University of Kansas, www.kumc.edu/instruction/medicine/anatomy/histoweb, has detailed photos of the processes in each of the body systems. Some are quite humorous as well.

Nurse Healer www.nursehealer.com, click the site map and then click the Anatomy and Physiology link under Education to find a list of some of the best sites on the Internet for Anatomy and Physiology.

Next time you're on the Internet, don't forget to drop by www.cliffsnotes.com. We created an online Resource Center that you can use today, tomorrow, and beyond.

Other Media

Principals of Anatomy and Physiology with CD, by Gerald J. Tortora and Sandra Reynolds Grabowski, contains a solid scientific presentation of the principles of human structure and function. Includes excellent color diagrams, examples of clinical applications and disorders, and flowcharts of feeback mechanisms. Wiley, John & Sons, Inc. $103.75.

anatomyLINX, an interactive multimedia electronic atlas of the human body, available on CD-ROM by Horizon Innovations. You can check out this software at www.horizoninnovations.com.

Dynamic Human 2.0, an interactive software package that features anatomy, clinical concepts, explorations, histology, and includes additional material online including maps, indexes, and quizzing on each body system. Ordering information is available at www.mhhe.com.

Glossary

active transport the movement of solutes against a gradient and requires the expenditure of energy (usually ATP).

anatomy the study of the structure and relationship between body parts.

antagonistic hormones hormones that act to return body conditions to within acceptable limits from opposite extremes.

antigen any molecule, usually a protein or polysaccharide, that can be identified as foreign or nonself. It may be a toxin, or a part of the protein coat of a virus, or a molecule unique to the plasma membranes.

appendicular skeleton the bones that make up the limbs, basically your arms, legs, hands, and feet.

arteries carry blood away from the heart.

atom the smallest quantity of an element that still possesses the characteristics of the element.

ATP (adenosine triphosphate) a common source of activation energy for metabolic reaction.

atria the two upper chambers of the heart.

autonomic nervous system (ANS) controls the activities of organs, glands, and various involuntary muscles.

axial skeleton the bones that revolve around the vertical axis of the skeleton, basically the spine and skull.

basal bodies at the base of each flagellum and cilium and appear to organize their development.

belly the part of the muscle between the origin and insertion.

Boyle's law describes the relationship between the pressure *(P)* and the volume *(V)* of a gas. The law states that if the volume increases, then the pressure must decrease (or vice versa).

capillaries microscopic blood vessels with extremely thin walls.

carbohydrate any of certain organic compounds, including the sugars, starches, and celluloses.

cardiac cycle describes all the activities of the heart through one complete heartbeat—that is, through one contraction and relaxation of both the atria and the ventricles.

cardiac muscle responsible for the rhythmic contractions of the heart.

cardiac output (CO) is the volume of blood pumped out of the right or left ventricle per minute. $CO = SV \times HR$.

cardiovascular system consists of the heart, blood vessels, and blood. This system transports nutrients, oxygen, and hormones to cells and removes metabolic waste. It protects the body and regulates body temperature.

catalyst accelerates the rate of a reaction.

cell the basic functional unit of all living things.

central nervous system (CNS) consists of the brain and the spinal cord.

centrioles a pair of centrioles located outside the nuclear envelope gives rise to the microtubules that make up the spindle apparatus used during cell division.

cerebrum consists of two cerebral hemispheres connected by a bundle of nerve fibers.

channel proteins provide passageways through the plasma membrane for certain water-soluble substances such as polar and charged molecules.

chromosomes made up of two identical halves called sister chromatids, which are joined at the centromere. Each consists of a single, tightly coiled molecule of DNA.

cilia structure that protrudes from the cell membrane and makes wavelike movements. Classified by length and number per cell, cilia are short and many. The numerous cilia that line the respiratory tract sweep away debris. Structurally cilia consists of microtubules arranged in a "9 + 2" array.

cofactors nonprotein molecules that assist enzymes.

compact bone the hard material that makes up the shaft of long bones and the outside surfaces of other bones.

compound when the atoms in a molecule are different. For example, H_2O and $C_6H_{12}O_6$, but not O_2.

cones photoreceptor cells that respond to bright light and color and transmit sharp images.

coronary circulation the blood vessels that supply oxygen and nutrients to the tissues of the heart.

covalent bond a chemical bond that forms when electrons are shared between atoms.

cranial nerves nerves of the peripheral nervous system which originate from or terminate in the brain.

cytokinesis divides the cytoplasm in the cell.

cytology the study of cells at the microscopic level.

cytoplasm specialized bodies within a cell called organelles suspended in a fluid matrix, the cytosol.

cytoskeleton the internal structure of the cytoplasm.

cytosol a fluid matrix of water and dissolved substances such as proteins and nutrients.

Dalton's law states that the sum of the partial pressures of each gas in a mixture is equal to the total pressure of the mixture.

dermis the second layer of skin.

dialysis the diffusion of solutes across a selectively permeable membrane.

diastole a relaxation event of either the atria or ventricles.

diencephalon connects the cerebrum to the brain stem.

diffusion the net movement of substances from an area of higher concentration to an area of lower concentration.

digestive system broken into two parts is responsible for the digestion and absorption of nutrients.

disaccharide a carbohydrate consisting of two linked sugar molecules.

DNA (deoxyribonucleic acid) the part of a cell that stores genetic information.

dorsal refers to the back of the body.

effector corrects conditions according to the evaluation of the integrator.

electron transfer proteins involved in transferring electrons from one molecule to another during chemical reactions.

electrons the negative charge of an atom outside the nucleus.

elements make up matter and possess unique physical and chemical properties.

endocrine glands secrete hormones directly into the bloodstream.

endocrine system produces hormones that are instrumental in maintaining homeostasis and regulating reproduction and development.

endoplasmic reticulum (ER) stacks of flattened sacs involved in the production of various materials.

enzymes proteins that are specific for particular reactions.

epidermis the top layer of skin.

epithalamus contains the pineal gland.

epithelial membrane a combination of epithelial and connective tissues working together to perform a specific function.

epithelial tissue acts as a covering and lining on the outside surfaces of the body and lines the internal organs or secretes hormones or other products.

exocrine glands secrete their substances into tubes, or ducts, which carry the secretions to the epithelial surface. Examples include sweat, saliva, milk, and stomach acid.

female reproductive system consists of the ovaries, uterine tubes, uterus (womb), vagina (birth canal), and vulvae.

flagella structure that protrudes from the cell membrane and makes wavelike movements. Classified by length and number per cell, flagella are long and few. A single flagellum propels sperm. Structurally flagella consists of microtubules arranged in a "9 + 2" array.

formed element cells and cell fragments present in blood.

frontal (coronal) plane divides the body or organ vertically into a front and rear part.

gametes eggs and sperm.

ganglia clusters of cell bodies in the peripheral nervous system.

gastrointestinal (GI) tract is a continuous tube with two openings, the mouth and the anus. It includes the mouth, pharynx, esophagus, stomach, small intestine, and large intestine.

Golgi apparatus (Golgi complex or Golgi body) a group of flattened sacs arranged like a stack of bowls that function to modify and package proteins and lipids into vesicles.

gross (macroscopic) anatomy the study of body parts visible to the naked eye.

gustatory sense sense of taste.

heart rate (HR) the number of heartbeats per minute.

hematopoiesis production of red blood cells.

hemopoiesis the process that produces the formed elements of the blood and takes place in the red bone marrow of long bones.

hemostasis the stoppage of bleeding through vascular spasm, platelet plug, or coagulation.

Henry's law the greater the partial pressure of a gas, the greater the diffusion of the gas into the liquid.

histology the study of tissues at the microscopic level.

homeostasis the maintenance of stable, internal conditions within specific limits. A characteristic of all living systems.

horizontal (transverse) plane divides the body or organ horizontally into a top and bottom part.

hormone chemical messenger produced by a cell that effects specific changes in the cellular activity of other cells.

hydrogen bond a weak chemical bond that forms between the positively charged hydrogen atom and one covalently bonded molecule and the negatively charges area of another covalently bonded molecule.

hypodermis the layer between the dermis and the underlying tissues and organs.

immune system the third line of defense that consists of specific mechanisms and agents that target specific antigens.

induced-fit model describes how enzymes work.

inorganic compound compounds that do not contain carbon atoms. Water, O_2, and NaCl are examples of inorganic compounds.

insertion the muscle end that attaches to the moving structure.

integrator a control center (often the brain).

integumentary system the skin and its accessory organs.

interatrial septum separates the left and right atria longitudinally in the heart.

interventricular septum separates the left and right ventricles longitudinally in the heart.

iris the colored portion of the eye that opens and closes to control the size of its circular opening, the pupil.

kidneys produce urine.

left atrium (and its auricle appendage) receive oxygenated blood from the lungs through four pulmonary veins and acts as a holding chamber for blood in readiness for its flow into the left ventricle.

left ventricle the pumping chamber of the heart for systemic circulation.

limbic system a network of neurons that extends over a wide range of areas of the brain. Imposes an emotional aspect to behaviors, experiences, and memories.

lipids a class of substances that are insoluble in water (and other polar solvents) but are soluble in nonpolar substances (like ether or chloroform).

living systems can be described from various perspectives, from the very broad (looking at the entire earth) to the very minute (individual atoms).

lymph nodes small, oval or bean-shaped bodies that occur along lymphatic vessels. They filter the lymph, destroy bacteria, toxins, and particulate matter, and they produce antibodies.

lymphatic system consists of lymphatic vessels, a fluid called lymph, lymph nodes, the thymus, and the spleen.

lymphocytes white blood cells (leukocytes) that provide an immune response that attacks specific kinds of nonself cells and foreign substances.

lysosomes vesicles from a Golgi apparatus that contain digestive enzymes that break down food, cellular debris, and foreign invaders such as bacteria.

macrophages enlarged monocytes (white blood cells) that engulf microbes and cellular debris.

major histocompatibility complex (MHC) (also called human leukocyte antigens, HLAs) is the mechanism by which the immune system is able to differentiate between self and nonself cells.

male reproductive system consists of the scrotum, testes, epidiymis, ductus deferens, ejaculatory ducts, urethra, spermatic cord, accessory sex glands, and the penis.

mammary glands sweat glands on a woman that are specialized for the production of milk.

matter anything that takes up space and has mass.

meiosis a reduction division, producing daughter cells that contain half the genetic information of the parent cell.

melatonin a hormone that helps regulate the biological clock (sleep-wake cycles).

metabolism the breakdown of substances, the formation of new products, or the transferring of energy from one substance to another.

microtubules protein fibers made of the protein tubulin and provide support and mobility for cellular activities found in the spindle apparatus.

mitochondria carry out aerobic respiration.

mitosis a process of cell division in which the nucleus is divided so that both daughter cells are genetically identical.

molecules atoms chemically bound together.

monosaccharide the simplest kind of carbohydrate consisting of a single sugar molecule.

motor neurons transmit impulses from the central nervous system to effectors (muscles or glands).

muscle tissues consist of skeletal muscle, cardiac muscle, and smooth muscle.

muscular system consists of skeletal muscles and their associated connective tissues. It does not include cardiac muscle or smooth muscle.

negative feedback a sensing mechanism detects a change in conditions beyond specific limits. A control center evaluates the change and activates the change and activates a second mechanism to correct the condition. The variant condition is canceled, or negated, so that conditions are returned to normal.

nephrons filtering units that number over a million in the kidneys.

nerve a bundle of nerve fibers in the peripheral nervous system held together with connective tissue.

nervous system consists of two parts: the central nervous system and the peripheral nervous system.

nervous tissue tissue that consists of two kinds of nerve cells, neurons and neuroglia.

neuroglia cells that support and protect neurons.

neuron a cell that transmits nerve impulses.

neurophysiology the study of how the nervous system functions.

neutrons the neutral charge in the nucleus of an atom.

nuclear division divides the genetic material in the nucleus.

nuclei clusters of cell bodies in the central nervous system.

nucleus the part of the cell that contains DNA, the hereditary information of the cell.

olfactory sense sense of smell.

oogenesis consists of the meiotic cell divisions that lead to the production of ova (eggs) in females.

organ a group of different kinds of tissues working together to perform a particular activity.

organelles bodies within the cytoplasm that serve to physically separate the various metabolic reactions that occur within cells.

organic compounds compounds that have carbon atoms.

organism a system possessing the characteristics of living things—the ability to obtain and process energy, the ability to respond to environmental changes, and the ability to reproduce.

osmosis the diffusion of water molecules across a selectively permeable membrane.

parasympathetic nervous system activates tranquil functions, such as stimulating the secretion of saliva or digestive enzymes into the stomach.

passive transport the movement of substances from regions of higher to lower concentration and do not require expenditure of energy.

peripheral nervous system (PNS) consists of nerves outside the central nervous system.

peristalsis the major means of propulsion in moving food along the digestive tract. Consists of a series of alternating contractions and relaxations of smooth muscle that lines the walls of the digestive organs and that forces food to move forward.

phagocytes white blood cells that wander throughout the body attacking bacteria and other foreign invaders.

physiology the study of the function of body parts and the body as a whole.

plasma (cell) membrane separates internal metabolic events from the external environment and controls the movement of materials into and out of the cell. It bounds the cell and encloses the nucleus and cytoplasm.

plasma straw-colored, liquid portion of the blood, which consists of water, proteins, waste products, nutrients, electrolytes, and respiratory gases.

platelets (thrombocytes) fragments of huge cells that adhere to damaged blood vessel walls and release enzymes that activate the stoppage of bleeding.

polysaccharide a carbohydrate that is a series of connected monosaccharides.

positive feedback an action intensifies a condition so that it is driven farther beyond normal limits.

primary sex organs (gonads) secrete hormones that produces gametes.

process of sight involves converting light energy into chemical energy.

proteins represent a class of molecules that have varied functions. Eggs, muscles, antibodies, silk, fingernails, and many hormones are partially or entirely proteins.

protons the positive charge in the nucleus of an atom.

pulmonary circulation deoxygenated blood travels from the right side of the heart to each of the two lungs. Within the lungs O_2 enters and CO_2 leaves so that oxygenated blood returns from the lungs to the left side of the heart.

pupil the circular opening of the eye. The size of the pupil is controlled by the iris. The size of the pupil regulates the amount of light entering the eye and helps bring objects into focus.

receptor a sensing mechanism.

receptor proteins provide binding sites for hormones or other trigger molecules.

red blood cells (erythrocytes or RBCs) transport oxygen and carbon dioxide in the blood.

remodeling the process of creating new bone and removing old bone.

reproduction the production of eggs and sperm and the processes leading to fertilization.

respiratory system delivers air to the lungs.

retina the inner nervous tunic of the eye that absorbs light in the pigmented epithelium and stores vitamin A by use of photoreceptor cells.

Rh blood group This is a complex group defined by the antigens produced by three different genes. The Rh factor determines positive or negative blood.

right atrium located in the upper right side of the heart, along with the right auricle, acts as a temporary storage chamber so that blood will be readily available for the right ventricle.

right ventricle the pumping chamber of the heart for pulmonary circulation.

RNA (ribonucleic acid) takes genetic instructions from DNA and directs various metabolic activities of the cell.

rods photoreceptors that are most sensitive to light and more numerous than cones. The provide vision in dim light and are more capable of detecting movement.

sagittal planes divide the body or organ vertically into a right and left part.

sensory neurons transmit impulses from skin and other sensory organ or from various places within the body to the central nervous system.

skeletal muscle attaches to bones and causes movements of the body; also known as striated muscle or voluntary muscle.

skeletal system bones, cartilage, and the membranes that line the bones.

smooth muscle lines the walls of hollow organs.

somatic nervous system (SNS) directs the contraction of skeletal muscles.

spermatogenesis the development of a spermatid into mature sperm cells.

spinal cord an extension of the brain stem that functions to transmit nerve impulses and spinal reflexes.

spleen the largest lymphatic organ and is located on the left side of the body between the diaphragm and stomach. It filters the blood, destroys old red blood cells, provides a reservoir of blood, is active in immune responses, and produces blood cells.

spongy bone consists of thin, irregularly shaped plates arranged in a latticelike network.

steroids are characterized by a backbone of four linked carbon rings. Examples of steroids include cholesterol and certain hormones, including testosterone and estrogen.

stroke volume (SV) the volume of blood ejected by each ventricle during a single contraction.

sympathetic nervous system involved in the stimulation of activities that prepare the body for action, such as increasing the heart rate.

synapse (synaptic cleft) a gap that separates the neuron from a muscle cell or another neuron.

systemic circulation oxygenated blood travels from the left side of the heart to the various areas of the body. Gas, nutrient, and waste exchange occurs across the capillary walls into the surrounding cells. The deoxygenated blood returns to the right side of the heart.

systole a contraction event of either the atria or ventricles.

tissue a group of similar cells performing a common function.

urinary system helps maintain homeostasis by regulating water balance and by removing harmful substances from the blood.

veins carry blood toward the heart.

ventricles the two lower chambers of the heart that serve as the pumping machines.

vesicles small, spherically shaped sacs that bud from the ends of a Golgi apparatus.

white blood cells (WBCs) protect the body from foreign microbes and toxins.

zygote the fusing of an egg and a sperm.

Index